D0064772

# Middle East Oil and U.S. Foreign Policy

# Shoshana Klebanoff

The Praeger Special Studies program—utilizing the most modern and efficient book production techniques and a selective worldwide distribution network—makes available to the academic, government, and business communities significant, timely research in U.S. and international economic, social, and political development.

HD 9576
.N36K57

# Middle East Oil and U.S. Foreign Policy

## With Special Reference to the U.S. Energy Crisis

PRAEGER SPECIAL STUDIES IN INTERNATIONAL ECONOMICS AND DEVELOPMENT

**Praeger Publishers**     New York     Washington     London

Library of Congress Cataloging in Publication Data

Klebanoff, Shoshana.
    Middle East oil and U. S. foreign policy.

    (Praeger special studies in international economics
and development)
    Bibliography: p.
    1.  Petroleum industry and trade—Near East.
2.  Energy policy—United States.   3.  United States
—Foreign relations.   I.  Title.
HD9576.N36K57      338.2'7'2820956      73-17727

INDIANA
PURDUE
LIBRARY
SEP 1974
WITHDRAWN
FORT WAYNE

PRAEGER PUBLISHERS
111 Fourth Avenue, New York, N.Y. 10003, U.S.A.
5, Cromwell Place, London SW7 2JL, England

Published in the United States of America in 1974
by Praeger Publishers, Inc.

*All rights reserved*

© 1974 by Praeger Publishers, Inc.

Printed in the United States of America

To my MOTHER and DAUGHTER

whose love sustained this work

No nation which lacks a sure supply of
liquid fuel can hope to maintain a position of leadership among the peoples of
the world.  It follows that if the United
States is to hold the place it now occupies
on the world stage as an effective leader
in elevating the standard of living for
people, it must develop a national petroleum policy which will make certain that
we shall not become dependent upon any
other country for our supply of liquid
fuel.

> Investigation of Petroleum Resources
> in Relation to the National Welfare.
> Final Report of the U.S. Senate
> Special Committee Investigating
> Petroleum Resources, 1947, pp. 1-2.

**PREFACE**

The underlying hypothesis of this study is the well-recognized economic fact that oil is the basis and the moving power of modern industrial society and is therefore indispensable. In order to bolster its economy, a nation must have a sound oil policy that links considerations of the availability of its domestic oil resources with external factors that affect world supply and demand in the international market. In order to secure its foreign sources of supply, a nation must articulate a foreign policy capable of protecting its interests in the producing areas. It also must depend on a military strategy capable of defending its pipelines on the land and its tanker fleets on the oceans.

When a nation's dependence upon foreign sources of oil becomes minimal, its domestic economy becomes to that extent free of the uncertain fluctuations of external influences. In times of tension, then, its foreign oil policy is likely to become preoccupied with considerations of bloc alliances and alignments. It will strive to deny a potential enemy access to the foreign oil, and it will do everything in its power to assure access to friends. Such a politically oriented foreign oil policy has a restrictive effect on the free flow of international trade, inasmuch as it limits the availability of foreign oil to a group (or groups) of nations. Within these political restrictions, the private sector still continues to formulate commercial oil policies according to its own best interests.

Other countries whose oil reserves are nationalized and subject to public policy are able to resort to such collective pacts, which in a free-enterprise society would be regarded as violating antitrust legislation. The result will be that the free interplay of market supply and demand no longer exists. It is replaced, instead, by state monopolies vying against each other in restrictive markets that they subdivide among themselves by international agreements and strive to "stabilize" through price-fixing pacts, forming in the process what amounts to "oil empires" run by national governments who have not only economic considerations but also political objectives.

In assessing the impact of these external influences on U.S. foreign policy, the present study focuses on the U.S. position in the international oil markets in terms of production, transportation, refining, and distribution, as well as the U.S. political and military stakes in the Middle East and their implications vis-à-vis Soviet trade expansion east of the Suez Canal.

The objectives of this political research are (1) to identify the specific U.S. government interest in Middle East oil in terms of the particular domestic needs of the United States and the U.S. government's relations with the U.S. international oil companies operating in the Middle East; (2) to assess the impact of Western Europe's dependence on Middle East oil and relate it to problems arising from U.S. policies in the developing countries, with special emphasis on those in the Middle East; and (3) to identify the contemporary imperatives for a revitalized oil policy in view of the changing conditions of nationalization, local rivalries, and Soviet penetration into the area.

This study attempts to combine an overall political analysis of the problem of Middle East oil with a discussion of the economic imperatives of the world market that rule its production, transportation, refining, and distribution. The frame of reference that is used as the guideline in modifying the boundaries of this research is U.S. foreign policy in its relation to Middle East oil. Although the research centers attention on the period between the Marshall Plan (1948) and the closure of the Suez Canal (1967), to confine it to this period altogether would miss one fundamental ingredient--namely, the exploratory analysis of the policy interplay between the U.S. government and the international petroleum companies operating in the Middle East. The question as to what role these companies actually play in shaping U.S. Middle East policy in general and its foreign oil policy in particular is very important in itself and bears the utmost relevance to the topic of this research. However, this role cannot be properly understood unless one attempts to trace the evolution of government-industry relations from their origin. Therefore, a section at the beginning of this study depicts the growth of public and private U.S. interests in Middle East oil from 1919 to 1947. These background observations are deemed necessary in order better to understand the role played by the international oil companies within the overall Middle East oil picture during the period under discussion.

The last part of this study concerns the current energy crisis in the United States and its relation to Middle

East oil. It analyzes the possible impact of Arab oil imports in view of the general U.S. economy, balance of payments, foreign policy implications, and overseas commitments.

In conclusion, I wish to extend my thanks to Professors Fred W. Neal and Harold W. Rood, whose comments have enriched this study, which originated as a dissertation in the Graduate Program in International Relations at Claremont Graduate School and University Center, Claremont, California. My thanks also to Professor John M. Sullivan, who has read the manuscript. Research for the entire study was done at the Library of Congress. I especially wish to thank the members of the research facilities staff of the Library for their excellent assistance, which exceeded at times the call of duty.

August 1973

Since the conclusion of this book some important events have taken place in the Middle East that bear out much of the thesis presented in these chapters. Particularly, the Arabs' use of oil as a political weapon (predicted in this book) has been brought into focus. The Soviet role in provoking this anti-Western policy invites a new examination of the meaning of détente as practiced by the Soviet Union and the role of oil in achieving its purposes.

December 1973

# CONTENTS

LIST OF TABLES

# LIST OF ABBREVIATIONS

| | |
|---|---|
| AIOC | Anglo-Iranian Oil Company |
| Aramco | Arabian American Oil Company |
| CED | Committee for Economic Development |
| CENTO | Central Treaty Organization |
| CPSU | Communist Party of the Soviet Union |
| EA | Economic Affairs Division (Department of State) |
| FPSC | Foreign Petroleum Supply Committee |
| HR | House Resolution |
| IPC | Iraq Petroleum Company |
| MEEC | Middle East Emergency Committee |
| NATO | North Atlantic Treaty Organization |
| NEP | New Economic Policy |
| NIOC | National Iranian Oil Company |
| NPC | National Petroleum Council |
| OAPEC | Organization of Arab Petroleum Exporting Countries |
| OECD | Organization for European Cooperation and Development |
| OEEC | Organization for European Economic Cooperation |
| OPEC | Organization of Petroleum Exporting Countries |
| SEATO | South East Asia Treaty Organization |
| UN | United Nations |
| USSR | Union of Soviet Socialist Republics |

# THE U.S. GOVERNMENT AND
# THE U.S. PETROLEUM INDUSTRY
# IN THE MIDDLE EAST

# 1

## BETWEEN WARS
## (1919-39)

U.S. Middle East policy cannot be discussed in isolation from private U.S. oil investments in that area.  These
investments were first made during the 1920s (Iraq) and the
1930s (Bahrein and Saudi Arabia).  The investors, however,
were not legally bound to comply with official State Department policies in that region.  They operated as independent
concerns, and, as such, their overseas activities were beyond the pale of any government decisions.  Thus, U.S. economic, political, and strategic actions as they have since
evolved in the Middle East were largely the result of the
interaction between public government policies and private
companies' activities in the area and the effect upon both
of the domestic and the international environments.  The
groundwork for much of the subsequent relations between government agencies and the international (Middle East) petroleum industry was established during the era between the two
world wars, at which time the overriding policy considerations were economic and commercial, complicated by the ob-
structionist behavior of rival British interests.

The first section of this chapter is an analysis of
incipient government-industry participation in Middle East
oil ventures.  It relates public foreign oil policy to the
domestic state of supply and demand and the principles of
the prevailing open-door doctrine.  It also explores the
commercial profit motivations behind the private companies'
actions, which led the way to the extensive operations subsequently conducted by U.S. capital investors in Middle East
oil.  The second section of this chapter links U.S. overseas
oil ventures with the reorganization of the executive drive
to bring the entire domestic oil industry under government
control; it also explains the failure of these attempts in
view of the U.S. system of checks and balances represented
here by the legal structure of a free enterprise economy.

3

DOMESTIC OIL CONSIDERATIONS AND FOREIGN OIL

According to a U.S. Senate report:

> Prior to 1920 American Companies either had
> been indifferent to foreign reserves or they
> had been largely frustrated in their efforts
> to acquire reserves in the Eastern Hemi-
> sphere, owing to restrictive national and
> colonial policies of foreign governments and
> of private oil-interests.  After 1920, how-
> ever, they became actively interested in
> foreign reserves, spurred by the dual fears
> of a prospective shortage of American oil
> and of a British-Dutch monopoly of foreign
> reserves.  This increasing interest was also
> stimulated by the prospective discoveries of
> large foreign reserves which would afford to
> their owners a ready supply of cheap oil,
> advantageously located to important markets.[1]

The U.S. petroleum industry was a victim of recurrent
oil "scares," notably in 1906, when there was a general belief that only a little more oil remained in the United
States.  This scare was dispelled in 1907, when oil was discovered in Oklahoma.  Subsequently, in 1920, W. S. Farish,
later president of Standard Oil Company of New Jersey, declared that "oil in Texas and Oklahoma was 'running out'."
In the same year, Dr. David White, chief geologist for the
United States Geological Survey, predicted the exhaustion
of U.S. oil deposits within 18 years, and the Department of
the Navy became concerned whether its naval fuel needs could
be supplied much longer by domestic resources.[2]  At the same
time, California's oil production was dropping and tankers
were being built to move oil from Mexico through the Panama
Canal.  But, for various economic and political reasons, the
Mexican oil output was then in a decline.[3]
    During the early 1920s the fear that U.S. oil reserves
were being exhausted prompted several of the largest U.S.
oil companies engaged in foreign trade to look for new oil
fields in Central and South America and the Middle East.
At the same time, World War I had left the great powers exceedingly oil conscious.  During those postwar years, it was
generally anticipated that the United States would be forced
either to reduce consumption or to import.[4]  The ensuing
fear that these foreign oil reserves upon which the U.S. industry might soon have to depend would fall under the dominant or exclusive control of another nation, moved Senator

Henry Cabot Lodge to warn Congress that "England was taking possession of the oil supply of the world."[5]   Alarm was quickly spreading among the oil companies that their efforts to acquire foreign concessions, especially in the Middle East, were fatally late.[6]  This situation, and a widespread fear of world oil shortage, persuaded the U.S. government to sponsor U.S. oil ventures abroad.  The Department of the Interior, which was always the major force behind all U.S. domestic and foreign oil policies, became greatly concerned with the future of domestic energy resources and its significance to the welfare of the nation. Persuaded by the general oil scare that had quickly spread from the petroleum companies to government circles and was augmented by the universal fear, the Department of the Interior charted a policy in support of U.S. oil ventures abroad and urged all government departments and agencies to assist them.

The known reserves in the Middle East until that time were in Iran, but there were already strong indications of considerable untapped oil reserves in the Mosul district of Iraq.  Standard Oil Company of New Jersey and some other major as well as independent U.S. oil companies became interested in these prospects and requested that the Department of State assist them to acquire exploration rights in that new oil field.

On the international scene, the San Remo Peace Conference (April 19-26, 1920) was then focusing worldwide attention on the future of the former territories of the Turkish Empire.  The mandatory power over Syria and Iraq was assigned at that conference to France and England respectively, pursuant to which an Anglo-French agreement was signed between the two powers on April 24-25, to be operative under the name of the Turkish Petroleum Company, a former Anglo-German oil enterprise that had acquired its exploration rights in this area from the Grand Vizir in Constantinople before the war.*  The British claim to the concessionary rights in Iraq was based on a communication, dated June 28, 1914, from Grand Vizir Said Halim Pasha to the Turkish

---

*According to an agreement signed by the governments of Great Britain and Germany in early 1914, German financial and oil interests were incorporated into the Turkish Petroleum Company, which, until that date, had been owned exclusively by the Anglo-Persian Company.  Ultimately, as a result of this agreement, the British company owned 50 percent of the proposed new concern, the Dutch-Shell Company owned 25 percent and the German group 25 percent.

Petroleum Company. The Turkish parliament did not ratify the grant. Nevertheless, after the war the British argued that their claim was none the less valid.

The U.S. government, however, did not consider this communication as "a definite and binding agreement to lease." The Department of State, alerted to protect U.S. claims for oil in that area, informed the British Foreign Office through the U.S. Ambassador in London that it was "unable to conclude that any concession was ever granted by the Turkish Government to the Turkish Petroleum Company" and invited the government of Great Britain to bring this claim for determination by suitable arbitration. Moreover, Acting Secretary of State Henry P. Fletcher informed the British Ambassador in Washington that the entire San Remo Petroleum Agreement between Great Britain and France was not recognized by the U.S. government as "applicable to the disposition of economic opportunities in mandate territories." However, the U.S. Ambassador in London, George Harvey, advised Lord Curzon that his government's chief point was that because of the U.S. contribution to "the victory over the Central Powers," there was no reason "to discriminate against the United States or to refuse to safeguard its equality of commercial opportunities in these territories."[7] In Washington, the Department of State viewed British and French intentions with suspicion. As early as 1919 the Department of State had already been advised by John Bassett Moore, who was then acting as an adviser to the Standard Oil Company of New Jersey in matters of international law, that France and England were planning to freeze U.S. oil interests out of the Middle East. The Department of State considered the San Remo Peace Conference agreement on oil as a proof of such a collusion.

Fearing that this division of territorial sovereignty under the mandate system might imply an exclusive control of the natural resources of these Arab lands, the U.S. oil industry began to clamor for an open-door policy. Moved by the initiative of Herbert Hoover, then Secretary of Commerce, the Department of State used the bargaining power of the U.S. oil industry to put pressures on the British government to recognize the U.S. claim for an equal share in the Mosul district oil. According to Herbert Feis,

> The American Government measured the relative American situation by comparing the current American rate of consumption with estimated reserves within the United States, and emerged with a dismal picture indeed. The British Government computed the position

by comparing current British consumption
with current production within the Empire,
and emerged with a modest position indeed.[8]

The numerical case for the United States was, then, that
while it possessed approximately 12 percent of the natural
petroleum resources of the world, its annual needs approxi-
mated 70 percent of the world's annual supply.[9] At the same
time, oil output in the British Empire was only about 2.5
percent of the world's production and, with the inclusion of
the British-owned Persian oil, it amounted to 4.5 percent.[10]

At the same time, the new nationalist government of
Turkey began to dispute the inclusion of the oil-rich province of Mosul in the mandate territory of Iraq. In the circumstances, it seemed possible that the U.S. government
would support the Turkish claim.* Thus, despite its unal-
tered conviction that the empire needed its own oil resources and that the Middle East was the place to have them,
the British government finally agreed to recognize the U.S.
demand for an equal share with the British, the French, and
the Dutch, in the Iraq oil venture. Sir John Codman of
the Anglo-Persian Oil Company (APOC) was sent to the United
States in 1921 and then again in 1922 to discuss the situation. In 1922 a provisional agreement was reached with the
U.S. oil group, according to which it would be allowed to
share 20-25 percent in the Turkish Petroleum Company. (For
all practical purposes, the U.S. group received the share
of the Deutsche Bank in the Turkish Petroleum Company.)
The exact status of the U.S. oil companies was to be determined at a later date. In September 1923, the company applied to the government of Iraq (now officially a British
mandate) for concessionary rights for oil explorations in
the provinces of Mosul, Baghdad, and Basra. On March 14,
1925 (after the League Council had awarded the Province of
Mosul to Iraq), the Iraqi government granted the Turkish
Petroleum Company a 75-year concession. The U.S. group,†

---

*The Second Lausanne Conference discussed the question
as to who was to have the oil of Mosul but was unable to arrive at a decision. Consequently, the peace treaty was si-
lent about Mosul oil, and the question was referred to the
League of Nations for determination. In 1925 the League
Council awarded the province to Iraq.

†The interested U.S. oil companies in the Iraqi venture initially were Standard Oil Company of New Jersey,
Standard Oil Company of New York (Socony), Vacuum Oil

registered as the Near Eastern Development Corporation, increased the pressure for a final agreement with the company, backing their demands with principles of the open-door policy as expounded by Secretary of State Charles Evans Hughes in November 1922. The Department of State supported the oil industry's claim for an equal share in the Iraqi oil on the grounds that such claims were "consistent with the principles underlying the open door policy of the government of the United States."[11]  Still in 1927, when the first oil output, totaling 338,000 barrels, was extracted from the Mosul district oil field, the status of the Near Eastern Development Corporation in the Turkish Petroleum Company had not yet been determined.  Not until July 31, 1928, was the final agreement, henceforth popularly known as the Red Line Agreement, signed between all the partners in the Turkish Petroleum Company.  According to it, each one of the four major partners--the Anglo-Persian Oil Company, Royal Dutch Shell Company, Compagnie Française des Pétroles, and the Near Eastern Development Corporation--received equal shares of 23.75 percent in the Turkish Petroleum Company.  The remaining 5 percent went to C. S. Glubenkian, who was the chief negotiator of the enterprise.  In addition, the Turkish Petroleum Company was granted exclusive rights to engage in oil production in Iraq.

These monopolistic rights disturbed President Calvin Coolidge and Secretary of State Hughes, who conceived them to be contrary to the open-door idea.  However, the U.S. group in the company was determined to maintain the agreement as it stood, or, it threatened, it would drop out of it, leaving all of the oil of Iraq to the British and French.  The threat was tenable on the grounds that in the mid-1920s the oil situation in the Western Hemisphere had changed, and, with large new discoveries in Texas and the Caribbean, scarcity had turned into plenty.  Under the circumstances, and since Europe was still largely fueled and energized by coal, there were no great profitable markets at the time awaiting Middle East oil.  However, in order to

---

Company, and Atlantic Refining, "each with a long tradition of foreign trade."  Other companies that had expressed an interest were Pure Oil, Cities Service, Texas Company, Gulf Oil Corporation, and Sinclair Oil Corporation.  Eventually, however, the U.S. investment participating in Iraqi oil included only Standard Oil Company of New Jersey and Socony-Vacuum Oil Company. (See Leonard M. Fanning, Foreign Oil and the Free World [New York:  McGraw-Hill, 1954], p. 47.)

prevent that oil from falling into the hands of its recent
wartime allies, U.S. diplomacy supported the agreement subject to its amendment by a sublease provision "that other
oil companies would be given a chance to share in the development of the oil fields of Iraq."[12]  This provision,
incorporated into the agreement in order to satisfy the
open-door principle, was never brought into effect.  In
1931, when the revised agreement was signed between the
government of Iraq and the company, the provision was omit-
ted altogether, and the attention of other U.S. oil interests was thus diverted in later years to the oil fields of
Saudi Arabia, Bahrein, and Kuwait.  In June 1929, the name
of the Turkish Petroleum Company was changed to Iraq Petroleum Company (IPC).  After nine long years of diplomatic
exchanges and negotiations, the first U.S. stakes were thus
put down in the Middle East.

The Iraqi concession, however, signified only the beginning of U.S. involvement in Middle East oil.  On December 21, 1928, Standard Oil Company of California bought for
a mere $50,000 from Eastern Gulf all rights under option
contract in Bahrein but declined to buy the same option
rights at al-Hassa and the Kuwait Neutral Zone, which Gulf
had offered Standard at the same time.*  The Bahrein concession, however, was contested by the British government,
which insisted that under its 1880 and 1892 treaties, no
concessions in that territory could go to foreign entities
unless approved first by the British Colonial Office, which,
on its part, insisted that only British companies be permitted to operate in Bahrein.  This hitch was overcome as
soon as Standard Oil of California proceeded, in 1929, to
register a wholly owned subsidiary, The Bahrein Petroleum
Company, Ltd., under the laws of Canada with domicile in
Ottawa.  (By the Balfour Declaration of the British Common-
wealth of Nations, a Canadian company would enjoy British
status.)[13]

The British government, faced with the evidence of its
own laws, had no choice but to approve the concession to
the Americans.  The Bahrein Petroleum Company then proceeded

---

*All three areas, Bahrein, al-Hassa, and the Kuwait
Neutral Zone, had a long and interesting history of British
exclusive option contracts acquired by Major Frank Holmes,
a New Zealander, for the British-owned Eastern and General
Syndicate.  Gulf had bought the rights from the syndicate
on November 6, 1927, "just a week before the British option
was due to lapse."  (See Christopher Tugendhat, Oil: The
Biggest Business [London:  Eyre and Spottiswoode, 1968],
pp. 88-89.)

to build a refinery, only eight miles from the producing
oil field, which became one of the largest in the Middle
East, with a producing capacity of 205,000 barrels per day.
However, the Bahrein Petroleum Company later sold half interest in its Bahrein concession to the Texas Company, and
the new merger, registered in the Bahamas, was named
California-Texas Oil Company (Caltex).

The money that California Standard had received from
this deal with the Texas Company it invested in a new oil
concession, now opened for the first time in Saudi Arabia
on the mainland. The development of oil exploration in the
al-Hassa region of Saudi Arabia between 1932 and 1938 led
directly to the formation of the Arabian American Oil Company (Aramco).

These three oil ventures, in Iraq, Bahrein, and the
al-Hassa region of Saudi Arabia, constituted the total of
U.S. oil investments in the Middle East between the two
world wars. In 1939, these investments amounted to only
10 percent of all Middle East oil production. The all-
important economic, political, and strategic role of Arabian oil did not become apparent until World War II made
the rapid development of oil production a major precondition for the Allies' victory. (See Table 1.)

TABLE 1

Middle East Oil Production, 1938
(barrels)

| Area | Non-American | American | Total |
|---|---|---|---|
| Iran | 78,372,000 | -- | 78,372,000 |
| Iraq | 24,950,000 | 7,693,000 | 32,643,000 |
| Bahrein | -- | 3,398,000 | 3,398,000 |
| al-Hassa | -- | 495,000 | 495,000 |
| Total | 103,322,000 | 11,586,000 | 114,908,000 |

Source:  Compiled by the author.


THE PETROLEUM INDUSTRY UNDER THE NEW DEAL

In a very short time, these U.S. oil companies operating abroad acquired a dominant position in foreign commerce, a situation that in turn had great impact in not
only the economic but also the political sphere.  These

companies maintained relations with governments of the producing countries as well as with governments of the consuming countries.  At the same time, they kept the Department
of State advised as to their current activities and programs "to the end that the same will not be contrary to
American foreign policy and that the companies will be in
a position to request prompt diplomatic protection when
necessary."[14]  The international role in which these companies were engaged as producers and suppliers of the
world's most sought-after commodity transformed the international petroleum industry from a purely economic concern
into a major agent of foreign policy, which, at the same
time, was independent of government control.  Indeed, the
concern of official circles about public oil policy was
viewed by these companies as running contrary to the companies' best interest.  At a time when concessionary rights
were still granted and honored, the companies repeatedly
insisted on their freedom from government intervention in
what they maintained was the private business of U.S. citizens abroad.

The increased influence of the international oil industry was matched by its already established status at
home.  The concentration of great economic power in the
hands of a few corporations spurred suspicions of practices
that were in violation of the Sherman Antitrust Act (1890).
During President Franklin Delano Roosevelt's first administration, these allegations soon became a political issue.
Although the public controversy between government and oil
men seemingly concerned only the home industry, its implications could have far-reaching effects on the policies of
the international oil companies as well.*  Any measure that
the federal government would decide to take in respect to
the home industry was going to determine the position of
the United States as a supplier and a consumer in the world
market.  Therefore, although the oil controversy, on the
surface, seemed to be confined solely to domestic oil, the
international oil companies could not help showing a very
keen interest in its development and expressed their concern at their International Oil Conference, which was con-
vening in Paris at that time.[15]

With the intention of coordinating production with
market demand,[16] the federal government attempted to assume

---

*Although all U.S. oil companies operating abroad had
also vast investments in domestic oil, their foreign and
domestic operations were always conducted as separate corporate entities.

control over both production and marketing, as well as the
setting of prices.[17]  Asking Congress to institute a comprehensive inquiry into industrial organization, President
Roosevelt said:

> The power of the few to manage the economic
> life of the Nation must be diffused among
> the many or be transferred to the public
> and its democratically responsible govern-
> ment.  If prices are to be managed and ad-
> ministered, if the Nation's business is to
> be allotted by plan and not by competition,
> that power should not be vested in any pri-
> vate group or cartel, however benevolent
> its professions profess to be.[18]

The recently enacted National Industrial Recovery Act
provided Roosevelt with the legal umbrella under which a
National Petroleum Code[19] was approved by him (August 19,
1933) and was handed over to Secretary of the Interior
Harold L. Ickes for administration (August 29, 1933).[20]
Self-government in industry, by majority rule, was the gist
of the National Recovery Administration.  To the small independent oil companies, this "self-government" implied the
threat of their being compelled to subordinate their interests to policies drawn up by the few major corporations
that controlled the largest segments of the industry.  Their
fear increased when, after a series of conferences between
the president and representatives of the major oil industries regarding the nature of operation of the code, the
latter announced that they had accepted the code and agreed
to abide by it.  For nearly two years, the "independents"
waged a futile battle against the code and against the "majors" within the restrictions imposed upon them by the code.
Within a very short time, it seemed evident that the restrictive nature of the code itself threatened to bring
forth a "lawless element," which, it was alleged, might
"imperil" the entire country.[21]  However, in 1935 the Supreme Court declared the National Industrial Recovery Act
unconstitutional, and thus the administration's intention
to bring the petroleum industry under its control had to be
dropped.  The Petroleum Code was replaced by the Interstate
Oil Compact of 1935.  But within a very short time, a campaign was launched that was aimed against the allegedly illegal practices of the industry and that culminated in a
charge by the Department of Justice that the oil industry

was, in fact, violating the Sherman Antitrust Act* in its price-fixing practices.

The international petroleum companies, having carefully observed and analyzed the domestic struggle for control of oil resources between government and private enterprises, immediately translated this conflict of public and private interests into a firm forewarning of a similar contest that might occur if and when their overseas assets became vital to the national interest. When such a controversy did indeed arise (as happened in the 1940s in respect to the Petroleum Reserves Corporation and the Anglo-American Agreement), the foreign operations committee of oil industrialists patterned its recommendations along the model of the Interstate Oil Compact of 1935. And, as in the case of the Petroleum Code, the failure of government to control private foreign enterprises was soon followed by an investigation by the Federal Trade Commission of alleged monopolistic activities by the international oil companies. (For detailed discussions of the Petroleum Reserves Corporation, the Anglo-American Oil Agreement, and the Federal Trade Commission's Report, see below Chapters 3 and 5.)

## SUMMARY AND CONCLUSIONS

After World War I, U.S. government agencies and private U.S. oil companies directed their attention to overseas sources of petroleum. Soon, their separate efforts converged upon Middle East oil. The reasons for the government's interest in Arab oil resources were twofold. In the first place, geological surveys and forecasts predicted national and worldwide shortages of oil. The U.S. economy, already converted to and motorized by oil to a large degree, stood to suffer more than any other economy from such shortages. The government, therefore, sought to assure a constant inflow of foreign oil that would, in due course, supplement the presumably diminishing domestic resources. Secondly, the realization of the great importance of oil to

---

*The general purpose of the law "was intended to restore competition where it had disappeared, preserve it where it still prevailed, and perpetuate it as a system for controlling the American economy" (George W. Stocking and Myron W. Watkins, Monopoly and Free Enterprise [New York: Greenwood Press, 1968], p. 256).

modern warfare, which stemmed from the recent battlefield experience, placed petroleum on the list of high-priority strategic commodities, as no nation could win a mechanized war without a sufficient supply of oil. Thus, foreign oil reserves assumed a strategic as well as an economic dimension.

Two reasons also prompted the major domestic oil companies to look for overseas reserves. In the first place, the same oil scare that brought about government actions in the Middle East also spurred these companies to look for foreign sources of oil reserves. Secondly, the production of Middle East oil cost considerably less than that of domestic oil and thereby increased the profit margin of the producing companies. Several conditions combined to reduce the production cost of Middle East oil: On the average, every fifth oil well drilled in the Middle East was a producing well, as compared to 1 in 13 wells in the United States; oil was found in the Middle East far closer to the land surface than elsewhere, thus reducing the drilling cost per well; the Middle East provided cheaper manpower; and, finally, the fine consistency of Middle East oil made its refining expenses less costly than those of crude oil derived from other parts of the world.

Commercial profit interest, then, motivated the private companies to look to Middle East oil, while national economic and strategic considerations spurred government activities. But while the U.S. government was limited to formal diplomatic negotiations in its relations with other governments, the private companies could deal with foreign governments as well as with private corporations. They could register subsidiary affiliates under the laws of other nations, or they could enter into agreements among themselves as well as with foreign concerns. At the same time, the international oil companies also availed themselves of the diplomatic services that were made available to them by the Department of State. Thus, they had a wider range of means at their disposal by which to make their inroads into Middle East oil development.

This type of relationship between government and companies was in full accord with traditional U.S. foreign policy. Two major principles had always guided this foreign policy: First, diplomatic relations maintained with foreign nations were undertaken principally in order to assist the commercial interests and money-making power of U.S. citizens abroad; and, second, U.S. foreign policy continuously shied away from any hint of embroilment in the political affairs of other peoples. This second principle was particularly enforced in regard to the Middle East, which the U.S. government considered to be a sphere of European interest. "In sum then," writes Raymond Hare, "the situation at the outbreak of World War II was that, whereas we had consistently and sometimes vigorously asserted American rights in the Middle East, we had deliberately kept our political profile low in what someone has described as a 'reverse Monroe Doctrine'."[22] U.S. diplomatic intervention in favor of economic interests between the two world wars was generally backed by the open-door doctrine, which, in principle, strove to extend the protection of free competition beyond the national territory. U.S. diplomacy, however, appeared to be more active in support of such overseas enterprises as were in the interest of the national welfare. Thus, the State Department's circular of August 6, 1919, read, "The vital importance of securing adequate supplies of mineral oil both for the present and future needs of the United States has been forcibly brought to the attention of the Department."[23] Although government-guided investments of U.S. capital were already being made in several Latin American countries, the Department of State assisted U.S. capital to get involved in Middle East oil much more than it did for any other foreign enterprises.[24]

On the domestic scene, at the same time, the Roosevelt Administration, acting under the umbrella of the National Industrial Recovery Act, attempted to subordinate the domestic petroleum industry to government control. In 1935, the Supreme Court declared this act unconstitutional. Nevertheless, while it lasted (1933-35), the National Recovery Administration, in its relation to petroleum, helped to focus attention on the fundamental conflict of interests between government and companies, stemming from the question of the power of control of the available domestic resources.

This struggle for dominance between centralized federal agencies and the private sector foreshadowed, in many respects, the ensuing rivalry between government and companies over Middle East oil, which came to a head during World War II and which will be discussed in the next chapter.

Thus, the interwar period already sharply defined the frame of references for much of the future relationships between the various U.S. actors converging upon the area of Middle East oil. The initial motive that had brought all these actors to the region was the consistent concern in the 1920s and 1930s about the diminishing domestic resources of petroleum. However, the immense importance of

this strategic commodity for national security and its
great commercial potential were not demonstrated in reality until World War II, during which Arab oil became the
vehicle for Allied victory.

# 2

## OIL FOR FIGHTING

World War II constituted a turning point in U.S. foreign policy in regard to Middle East oil.  Up until that
time, U.S. administrations had fostered diplomatic relations
with nations of this region only in order to be in a position to protect traditional U.S. commercial and missionary
interests abroad.  However, the immense contribution of
Middle East oil to the war effort changed the oil's intrin-
sic value from that of a primarily commercial asset to that
of a high-priority strategic commodity.  Thus, the national
interest itself became interwoven with the interests of private enterprise.

The first section of this chapter concerns the growing
dependence of the war effort on ready supplies of oil.  The
second section analyzes government attempts to preempt the
managerial control over Saudi oil from the private companies and to place it under the jurisdiction of a public
agency.  This policy put the U.S. government in the role of
a competitor rather than a protector of the international
oil companies.  This fundamental change in U.S. Middle East
oil policy orientation repeated, in certain essential per-
spectives, the government's experience with the National
Industrial Recovery Act under the New Deal.  This policy
will be further analyzed in greater depth in the summary
section at the end of the chapter.

OIL AND STRATEGY

On December 2, 1942, Secretary of the Interior Ickes
was made Petroleum Administrator for War.*  His immediate

---

*On May 28, 1941, the prewar armaments program led
President Roosevelt to set up the office of Petroleum

17

task was to attend to the U.S. oil problem, which stemmed from shortages in both supply and transportation. High military consumption, caused by battlefield demand, drained the oil reserves of the United States, and "the withdrawal for military service of tankers which normally supplied the East Coast and Pacific Northwest"[1] made adequate distribution of oil from other Western Hemisphere sources hard to manage. To meet the challenge of supply, Ickes, in 1943, enforced the emergency laws by rationing all oil products in the United States. To meet the challenge of transportation, U.S. shipyards soon began to construct a large number of tankers, which began "to move supplies to the theatre of action in Europe, Africa and the Far East."[2] Working under the double handicap of shortages in materiel and manpower, the war's "unsung heroes," the tanker crews, began to move oil in every direction. The international petroleum industry itself was harnessed to the national war effort. The Middle East refineries "made available to Allied Armed Forces large quantities of Navy Special, Diesel Oil, Avgas [aviation gas], and Mogas [motor gas]."[3] The refineries supplied "not only the Allied armies in the Middle East and the Eastern Mediterranean, but also the Allied air forces, naval squadrons and land armies in the Indian Ocean area, and in Western Australia."[4] They provided fuel for 185,000 U.S. bombers and fighting planes, 18 million tons of merchant ships, 120,000 tanks, and hundreds of thousands of trucks and other vehicles that carried supplies for the Allied forces and for use at home.

Meantime, at the Petroleum Administration in Washington, the Foreign Division did much of the spade work for the oil industrialists. This work involved, among a great deal more, the study of the geography of military operations. It required "a constant study of the future probabilities so that tankers and steel might be saved by producing oil for war requirements for any particular operation at the most favorable spot."[5]

The wartime teamwork between government and oil industry produced a high degree of cooperation. It brought forth accolades from those who, in peacetime, were among the most severe critics of the oil industry's practices. However, this enthusiasm for the great contribution of big business to the national war effort was indeed short-lived. Very

---

Coordinator for National Defense. On December 2, 1942, Roosevelt broadened the duties of the coordinator by creating the Petroleum Administration for War.

soon the immense importance of petroleum as a strategic commodity and as marketable goods brought companies and government into conflict again over the question of its control
not only at home but also in the Middle East.

## THE SECRETARY OF THE INTERIOR
## AND MIDDLE EAST OIL

Secretary Ickes's experience as Petroleum Administrator for War spurred his ambitions to have the international
petroleum industry permanently put under government control.
His imagination was captured by the immense possibilities
of such an enterprise and their application to the vital interests of the nation.  In particular, he set his sights on
Middle East oil, and, more precisely, he coveted the oil of
Saudi Arabia.  The grand scale with which Aramco was able
to respond to the national emergency impressed him so much
that he began to contemplate the acquisition by the U.S.
government of these oil concessions as a legitimate government business.*

Ickes was not alone in his enthusiasm for government
control of foreign oil resources.  As the war progressed,
the conviction grew within official U.S. circles that the
government itself must play an active part in assuring that
the United States or, at least, U.S. interests would control foreign oil resources adequate for the country's future
needs.  As noted above, the Department of the Interior became increasingly concerned about the possible exhaustion of
domestic and nearby reserves and about the necessity of sup-
plementing them by foreign production; in addition, the Department of the Navy desired that foreign reserves for its
exclusive use should be acquired and clearly set aside under
government control (such reserves had already been established within the United States and were sought in other
parts of the world); and the Department of State's adviser
for International Economic Affairs, Herbert Feis, was engaged in defining goals and guiding principles for a U.S.

---

*According to Ambassador Raymond A. Hare, the stronger
line advocating full government control of Saudi oil was
supported by Ickes, while Secretary of State Cordell Hull
favored a softer line, such as contracting for output.  See
Raymond A. Hare, "The Great Divide:  World War II," Annals
of the American Academy of Political and Social Science 401
(May 1972): 27-28.

foreign oil policy. Feis summed up the department's view of the main elements in the situation as follows:

1. U.S. "natural" production would probably not be sufficiently independent to meet U.S. requirements, even in peacetime.

2. "Synthetic" petroleum products could be manufactured in the United States, in whatever quantities might be required of them, but at heavy expense.

3. Foreign supplies of natural petroleum would be available for import into the United States or to satisfy foreign demands hitherto supplied from the United States.

4. U.S. reserves were a smaller fraction of world reserves than U.S. production was of world output, and the prospect of vast new discoveries was far better in certain other parts of the world than in the United States.

5. The satisfaction of possible future military requirements involved considerations not only of sufficiency and cost of supply but also of location and certainly of military control.[6]

Current developments seemed to open the door for active government involvement in the affairs of Middle East petroleum. In 1940 and 1941 King Ibn Saud was asking for a further advance of $6 million annually for five years, in addition to the lump sums of $10 million and $12 million that Aramco had already advanced him. In order to meet the king's threatening pressures to transfer concessionary rights to British companies if his demands were not met, Aramco devised two main schemes. One was a lend-lease program to Ibn Saud (see below), and the other one was a form of U.S. government ownership of Saudi Arabian oil. On April 16, 1941, the company proposed to President Roosevelt that

> the United States Government purchase from the Saudi Arabian Government finished petroleum products--gasoline, diesel and fuel oils--at very low prices, to the value of $6,000,000 annually for a period of six years. The Company would contract with Ibn Saud to produce, manufacture, and load such products for his account at a Persian Gulf port, while the King on his part would waive the royalty on an amount of crude oil corresponding, in current royalty rates, to $6,000,000 annually.[7]

This plan was brought to the attention of the Senate committee investigating Saudi Arabian petroleum resources.

Accordingly, it was stated at those hearings that "The purpose behind this arrangement between the Government of the United States and Saudi Arabia was to involve the United States Government in the Saudi Arabian concession instead of having the United States buy directly from the company; the King would then not be constantly threatening the Company."[8]

Thus, once again the separate interests of government and companies in foreign sources of petroleum converged upon the Middle East. On one hand, Aramco was charting plans for U.S. involvement in Saudi petroleum by way of direct purchases from the Arab monarch. On the other hand, various government agencies had expressed interest in a managerial concession of the same oil reserves. However, with the exception of proposals by Ickes, no such pragmatic scheme as devised by Aramco had emerged from their efforts. Both the Department of State and the Navy Department were largely engaged in general policy assessments of Middle East oil resources and their relation to the national interest of the United States, but they produced no applicable plan.

At the same time, at the Department of the Interior, Ickes was becoming increasingly committed to the proposition that the U.S. government ought to step into the picture not as a mere customer of an abundantly available petroleum supply, as suggested in Aramco's plan, but as an outright owner of these resources. Nothing less than a direct ownership of such oil supplies by the U.S. government itself could provide their security for the national interest. To the end that the U.S. government would be in a better position to acquire concessionary rights over Arabian oil, Ickes initiated a lend-lease agreement with King Ibn Saud, as had originally been conceived by Aramco.

On February 18, 1943, Roosevelt authorized Undersecretary of State Edward Riley Stettinius, then Lend Lease Administrator, "to arrange for lend-lease aid to the Government of Saudi Arabia," whose defense, the president hereby determined, was "vital for the defense of the United States."[9] In the spring of that year Brigadier General Patrick J. Hurley went to Saudi Arabia as Roosevelt's representative to investigate the U.S. oil position there and to arrange for a state visit from that country. Early in October, King Ibn Saud's son paid an official visit to Washington and remained in the United States a whole month. The lend-lease agreement with Saudi Arabia thus appeared to be sealed with ties of friendship between the two countries.

21

Having accomplished a lend-lease agreement with King
Ibn Saud, Ickes now approached the president with a proposition that the United States should take "immediate action
to acquire a proprietary and managerial interest in foreign
petroleum reserves."[10] He then broached his plan, in "emu-
lation of the government owned Anglo-Iranian Oil Company,"[11]
for the formation of a government-owned Petroleum Reserves
Corporation, whose task, he envisioned, would be to extract
oil from Middle East wells under U.S. government management.
"The first order of business of the Corporation," Ickes explained in a letter to the president, dated June 10, 1943,

> should be the acquisition and managerial in-
> terest in the crude oil concessions now held
> in Saudi Arabia by an American company.  The
> potential crude oil reserves underlying this
> concession have been estimated at approxi-
> mately 20,000,000,000 barrels--as large as
> the total current known crude oil reserves of
> the entire United States.
>      Their acquisition will serve to meet an
> immediate demand by the Army and Navy for
> large volumes of petroleum products in or
> near Arabia and will also serve to counter-
> act certain known activities of a foreign
> power which presently are jeopardizing
> American interests in Arabian oil reserves.[12]

The "foreign power" referred to is Great Britain, which
was, in the U.S. view, trying to oust U.S. companies from
Middle East oil enterprises.

On June 30 Roosevelt established the Petroleum Reserves Corporation as a last-minute action by the
subsidiary-creating powers of the Reconstruction Finance
Corporation, which was terminated on July 1 and transferred
to the Foreign Economic Administration headed by Leo T.
Crowley.  Ickes was made president of the new corporation
and chairman of its board of directors, which included Secretary of State Cordell Hull, Secretary of War Henry L.
Stimson, Secretary of the Navy W. Franklin Knox, and Crowley.  The purpose of the new agency was described as the
acquisition of "petroleum, petroleum products and petroleum reserves outside the continental United States."[13]

With his usual energy, Ickes immediately directed the
attention of the Petroleum Reserves Corporation to the Middle East.  The corporation's first act was to send a group
of geologists to survey the Middle East to report on oil

reserves, development, and potentialities. Next, the corporation, in August 1943, started to negotiate with Aramco
for the purchase of the latter's entire stock, thus marking
an important departure in U.S. foreign economic policy,
since it would have been the first time that the United
States would have owned foreign oil properties.

Roosevelt showed great interest in the plan as es-
poused by Ickes. The scheme also made great waves in all
government departments. It was described by some officials
as the Interior secretary's own project. Others, competing
for the credit of having conceived it, attributed it to the
secretary of War, while still others, apparently for similar reasons, credited it to the secretary of the Navy.
Whichever secretary's baby this idea was, all three departments gave the plan their strong support.

But Aramco, it was now disclosed, was not interested.
While foreign oil industrialists wanted the federal government to assume direct diplomatic responsibilities for U.S.
Middle East oil enterprises, they did not want to see it
own any part of them.

In the meantime, in November of that year, the group
of geologists sent its first report from the Middle East,
which confirmed that

> the center of gravity of world oil produc-
> tion was shifting from the Gulf of Mexico-
> Caribbean area (in which major United States
> fields were included) to the Persian Gulf
> area and was likely to shift until it was
> firmly established there. . . . In any at-
> tempted rating of undrilled prospect values,
> Iran and Saudi Arabia must vie for first
> place.[14]

In view of this report, and failing the Aramco deal,
the enterprising secretary of the Interior next "canvassed
ways and means to extend the United States Government control over the promising prospects of Kuwait, a concession
jointly held by Anglo-Persian Oil Company, Ltd., and Gulf
Oil Corporation."[15] Again, the concessionaries were not interested in the proposition of parting with their assets and
advised Washington to that effect. All these oil companies,
incorporated in Aramco and in Kuwait, also turned down a
modified offer by Ickes to buy only a "participating share"
in their stock.

Thus, it became evident that the primary purpose for
which the Petroleum Reserves Corporation had been formed,

namely, the direct control by the U.S. government of Middle
East oil, was unattainable.  Ickes then devised a new plan
according to which he proposed that the U.S. government own
and operate in conjunction with a private oil company an
oil refinery in the Middle East.

> The question of the refining capacity of
> Saudi Arabia for immediate military needs
> was being seriously discussed by the inter-
> ested authorities even before the organiza-
> tion of the Petroleum Reserves Corporation.
> . . . When it became obvious that the Gov-
> ernment would not acquire direct ownership
> of the Saudi concession, it was proposed that
> it build a refinery of 100,000-150,000 bar-
> rels' capacity; at that time the oil com-
> panies declared the proposal impractical.[16]

A government-owned refinery in the Middle East seemed
to pose a double jeopardy to the international petroleum industry:  It would not only be competing with the companies'
own refineries, but such a refinery would probably enjoy
greater official support than that extended to private
enterprises in that region and all military purchasing
of oil products would then be diverted from the companies'
refineries, which thus would lose their most lucrative single customer.  In effect, however, the progress of the war
at that time appears to have constituted the determining
factor in ending the negotiations regarding the refinery
project, since it could not be completed within one year.[17]
    Thus, with the exception of sending some geologists to
the Middle East, the Petroleum Reserves Corporation failed
to materialize any of its plans for the ownership by the
government of foreign oil reserves.  The corporation failed
to play the role it had intended to play in foreign oil developments.  Therefore, in the middle of January 1944, a
resolution was introduced in the Senate providing for the
liquidation of the Petroleum Reserves Corporation.  However,
the agency was temporarily salvaged from immediate dissolution by still another plan proposed by Ickes for the government to get a foothold in Middle East oil enterprises.
    In this plan, made public on February 6, 1944, Ickes
envisaged a government-owned network of pipelines throughout
the Middle East to transport crude oil from Bahrein, Saudi
Arabia, and, eventually, when discovery was made, also
Kuwait to a port on the Eastern Mediterranean.  The exact
route was not indicated; some reports suggested Haifa as

the probable port of destination, while others indicated the
Suez Canal or Alexandria. Ickes estimated the initial cost
of the pipeline to be between $130 million and $165 million.[18]
Again, the decision to build this pipeline was explained on
the grounds of its being an economic necessity in order to
conserve the oil supplies of the United States, which were
said to be fast running out. However, Ickes also stated
that the main objective of the new construction would be one
of strategy, as it would "greatly help to assure an adequate
supply of petroleum for the military and naval needs of the
United States in view of the obligations which this country
must assume for the maintenance of collective security in
the post-war world."[19] The "adequate supplies" required,
suggested Herbert Feis, would be obtainable through the ownership of such a pipeline, "as it would place the American
Government in a position to induce or possibly, to some extent, even to compel the companies to extend production if
that was desired."[20] Thus, for the first time, government
concern about future U.S. political involvement in Middle
East affairs entered considerations of oil policies.

Such a plan was more amenable to the Middle East companies. Transportation of oil had always been their thorni-
est marketing problem. As a rule, they did not own more
than one-half of their total tanker requirements and assigned the rest of their oil to tanker fleets of individual
owners. Pipelines posed a greater difficulty inasmuch as
they traversed national territories immersed in conflicting
and often hostile politics, with which none of the oil companies was properly geared to deal. It seemed to the U.S.
oil companies operating in the Middle East that if the U.S.
government desired to go into the pipeline business, so
much the better for their own purpose of ejecting themselves
from local politics. Moreover, they might have reasoned,
the government had all the ways and means at its disposal to
deal with political contingencies, which the companies did
not possess, and, in being involved in problems of oil profits, the government would soon find itself in a position
where it would become directly interested in protecting the
producing companies, an assurance that the international petroleum industry had been unsuccessfully trying to receive
from the Department of State since 1920. Consequently, on
February 6, 1944, an agreement was concluded between the
secretary of the Interior and the presidents of Standard Oil
Company of California and the Texas Company (the two coowner companies of Aramco), according to which the U.S. government undertook to "construct and to own and maintain a
main trunk pipe line system, including requisite facilities

for the transportation of crude petroleum from a point near the presently discovered oil fields of Saudi Arabia and Kuwait to a port at the eastern end of the Mediterranean Sea."[21]

This pipeline plan, as reported in the press, was blown out of all proportion to the realities of the enterprise. Thus, the New York *Times* carried a front-page story that stated that the Petroleum Reserves Corporation not only was to finance a vast network of pipelines throughout the Middle East but would also contribute funds necessary to finance the expansion of existing refineries and the construction of new ones in the Middle East by U.S. companies at a combined increased capacity of 600,000 barrels a day. This gigantic project was heralded as "the biggest single development in the history of the oil industry." (The story was reported by J. H. Carmical, the petroleum correspondent of the *Times*. Carmical also noted the difficulties of the project: "In view of the shortage of ocean shipping, the task of moving the material from this country, from where virtually all of it will have to come, will be difficult. Based on past experience, it is estimated that around 1,250,000 tons of steel alone will be required.")[22] The cost of the project was placed at several hundred million dollars, and the time required to complete it was estimated at 18 months to two years. The plan called for the expansion of the existing Haifa refinery to 350,000 barrels a day capacity, and the construction of a new refinery at Alexandria, with a capacity of 250,000 barrels a day. Alexandria was chosen because of its strategic location.

The objective of this vast oil enterprise, according to the New York *Times*, most significantly for future development of U.S. foreign oil policy in the Middle East, was "to increase the amount of oil available to the European area." However, because of the time lapse required for its completion, this project was considered more in terms of postwar requirements. This was the first time in U.S. Middle East policy that oil reserves in that region were said to be sought not in order to supplement diminishing U.S. reserves, but rather in order to bolster the European economies. This view had anticipated U.S. concern with Western Europe's postwar recovery and subsequent U.S. aid to the continent during two decades of oil crises.

The news of the government's intention to invest in a Middle East network of pipelines aroused a great deal of excitement in the country. There were many speculations as to the real purpose of the project. What was initially intended by Ickes to be a solution to strategic and economic problems

that vexed the Allies, and most particularly the United
States, soon became a matter of many contradicting political considerations. The press expressed anxiety not only
over the prospects of "the dwindling of oil reserves in the
United States," which the Department of the Interior had
set out to protect by turning to Saùdi oil reserves, but
also about the prospects that "the undertaking of this enterprise would draw the American government into the center
of all Middle Eastern political affairs" and that "this pipeline would likewise make this government a significant factor in the petroleum affairs not only of the Middle East but
of the world."[23]

Reactions to the projected pipeline, as reported in the
press, varied from group to group. Liberals, reported the
Economist of London, denounced it "as a further step in the
cartelisation of the oil industry, stressing the status of
the government as a privileged buyer rather than as controlling operator of the pipeline."[24] Herbert Feis observed
that in liberal opinion the pipeline was "disturbing, even
sinister, as a glistening corridor for imperialists."[25] On
the other hand, conservatives viewed government intervention
in private enterprise overseas with growing alarm. It signified to them a repetition of the same struggle for the
power of control of domestic resources that the country had
experienced in the early years of the New Deal era. At the
same time, isolationists feared that the projected pipeline
would "draw the United States into the historic struggle between the British Empire and the U.S.S.R. in the Persian
Gulf region."[26]

> But the fate of the project was not decided
> by such misgiving in regard to its politi-
> cal implications. It was rejected mainly
> because of the opposition of groups who be-
> lieved it would have an adverse effect upon
> their business position or prospect. Vir-
> tually the whole of the American oil indus-
> try condemned the proposed measure as need-
> less and unfair.[27]

The part of the U.S. oil industry that aligned itself
with the opposition to this project included the independent
domestic companies as well as the international oil companies
operating in fields other than those of the Middle East.
Both groups feared that a government stake in Middle East oil
would put the U.S. government in a position of a competitor
of other, non-Middle East oil companies and would induce it

to show unfair favoritism to those oil companies that, by
virtue of this investment, would now become its business
associates. The government would thus develop a large
vested interest in their welfare to the disadvantage of
other oil companies.

On March 2, 1944, therefore, the Petroleum Industry
War Council, consisting of 55 oil companies, but in this instance not including the three major companies engaged in
Arabian oil (Standard Oil Company of California and Texas
Company in Saudi Arabia and Bahrein, and Gulf Oil Company
in Kuwait), objected to the project.[28] The council registered its objections in a "white paper" prepared by George
A. Hill, Jr., of Houston, vice president of Independent Petroleum Association of America and a member of its national
oil policy committee as well as of the Petroleum Industry
Working Committee, an advisory group created by Ickes in order to sound out the oil industry and get its comments regarding the anticipated diplomatic negotiations with the
British government in regard to Middle East oil. (See Chapter 3, "A Negotiated Agreement.") The council's "white
paper" accused the Petroleum Reserves Corporation of secret
maneuvering, which had "provoked widespread discussions and
speculations." The objections to the proposed pipeline
were based on three grounds: economic, political, and strategic. The paper argued that since the project was more a
postwar plan than a wartime necessity, there was no justification for government-owned enterprises in what had always
been the free sector of the economy. Moreover, it was an
inter-Allied dispute involving the postwar world that was
being taken up in the midst of the present war to the detriment of the general war effort. The "white paper" accused
the U.S. government of using "the fascist approach" to
scuttle American free enterprise: "The fascist approach--
the corporate state, with its lust for imperialism--the
shackling of American free enterprise, not to better serve
but to displace private enterprise at not only the risk, but
the certainty, of international political involvement."[29]

The "white paper" also pointed out that, militarily
speaking, involvement by the U.S. government in Middle East
oil would burden it with immense security obligations for
which there was no historical precedent in U.S. relations
with that region. Involvement would raise

> the question of whether there should not be
> naval bases in the Persian Gulf, the Red Sea
> and the Mediterranean, air bases in the in-
> tervening areas and military bases and de-
> pots. . . . In this way only can the matter

be viewed realistically, and our past history affords no implied warrants for the
projection of this character of post-war
military policy with that degree of certainty that seemingly justifies present
commitments.[30]

In spite of the great secrecy that surrounded the construction of the proposed pipeline by the Petroleum Reserves Corporation, the oil companies were convinced that
the project signified "an entire new American foreign policy respecting petroleum."[31] They argued that this foreign
policy was being prepared by the Petroleum Reserves Corporation, an agency that was neither authorized nor equipped to
deal with matters of external politics. Moreover, the main
purpose of this agency, the oil industrialists maintained,
was not political but economic: It was meant to acquire
government ownership of oil properties, a step that would,
in time, involve the United States in future wars. James
A. Moffet, formerly Housing administrator in the Roosevelt
Administration, called the creation of the Petroleum Reserves Corporation "an unjustified adventure in bureaucracy"
and the worst "scandal" in the history of the oil industry.[32]
Moreover, he maintained that the pipeline idea itself was
commercially unsound, since it would be cheaper to move Arabian oil by tankers through the Suez Canal. If, however,
the government insisted on the necessity for such a network
of pipelines and refineries, it seemed to him that the private U.S. companies already holding concessions in the Middle
East were better equipped to deal with the project than a
government agency full of desk-bound bureaucrats. With only
sufficient diplomatic backing from the Department of State,
the private companies could develop Middle East oil enterprises without any financial assistance from the government.[33]
This upsurge of resentment against the projected pipeline prompted the Senate to form a special committee to
study the problem of oil resources in respect to U.S. national policy. Very soon, "the prevailing judgment of the
Senate became evident." The Senate was thoroughly against
the agreement, "and a vigorous group of Senators seemed determined not only to prevent its consummation but to compel
the dissolution of the Petroleum Reserves Corporation."[34]
(Four members of the Special Committee, representing interests of oil-producing states, led the battle in the Senate
against government ownership of oil properties, against intervention in private oil enterprises, and against the increase in power of the Middle East companies at the expense

of the domestic industry.  These senators were Senator Tom
Connally of Texas, Senator E. H. Moore of Oklahoma, Senator
Joseph C. O'Mahoney of Wyoming, and Senator John O. Overton
of Louisiana.)  By the time the Senate Special Committee had
completed its study, the pipeline was a dead issue.  The
findings and recommendations of the Senate Special Committee
Investigating Petroleum Resources are explored in greater
detail in Chapter 3.

On June 20, 1944, an Associated Press dispatch announced that Ickes's proposed 1,250-mile Arabian pipeline
had definitely been abandoned.  The government, it seemed,
had been pressed to leave all Middle East oil enterprises
to the private sector and, instead, now confined its attention, through the diplomatic channels of the Department of
State, to furthering U.S. interests in Middle East oil by
means of intergovernmental negotiations.


## SUMMARY AND CONCLUSIONS

As a result of the World War II experience, two essential factors in respect to oil emerged.  In the first place,
the subordination of the conflict of interests between government and the companies to the exigencies of the national
emergency was temporarily resolved in the placing of greater
centralized power of control in the hands of the federal government over domestic private oil interests.  This power of
control was manifested in the regulation of oil production
and in marketing and price ceilings and culminated in the
rationing orders that were issued to private consumers as
well as military personnel.  At the basis of these enlarged
government powers was the recognition that a vital commodity
serving the national interest ought to be managed by public
representatives and not by profit-motivated concerns.  This
was in fact the same principle that had underlined the National Petroleum Code of the previous decade.  However, the
power of control over private enterprise, which had been denied the government by Supreme Court decision in time of
peace, was not challenged in time of war.  There was a tacit
agreement that guaranteed individual freedoms were justifi-
ably repressible when serving the paramount goal of communal
survival, and the emergency laws enacted during that war un-
derscored this view.  In accepting this principle, the U.S.
government was in fact saying that the national interest
could not be entrusted to the hands of private citizens.

The second factor concerned the problem of sea-borne
transportation.  Because of pressing wartime military

requirements for a large naval construction, the commercial
end of such a build-up was lost sight of. Wartime tankers
were hastily constructed and were not meant to last very
long or to compete with the new modern tanker fleets of the
postwar era. During the war, the U.S. concept of a viable
merchant marine focused on its strategic uses and did not
attempt to adapt it to peacetime requirements.

At the same time as private oil concerns at home were
being subordinated to tighter government controls, increasing explorations in Middle East countries were yielding to
the Allies an enlarged production at smaller cost. The
emergence of the strategic role of Arab oil and its relation
to national security spurred the U.S. government to attempt
to expand its managerial power into these fields. A conflict of interests thus ensued between companies and government, calling to mind the controversy that followed the enactment of the Petroleum Code in 1933. While national interest was a paramount concern in Washington in relation to
Middle East oil, the international companies were primarily
interested in their own investments and future prospects and
were trying to protect both from government intervention.

In order to induce the Arab monarch to grant those concessionary rights to the U.S. government, Roosevelt, on February 18, 1943, signed a lend-lease agreement with Saudi
Arabia. Up until that time, lend-lease loans had only been
extended to war allies. The decision to include Saudi Arabia
in this scheme constituted a departure not only from lend-
lease loan policy but also from traditional U.S. foreign
policy in the Middle East.

For a century and a half U.S. foreign policy in the
Middle East had been concerned only with the protection of
the private interests of its citizens in this region. The
lend-lease agreement with Ibn Saud was not intended to extend protection to any private interests. It was purported
instead to promote a business association between the two
signatory governments. Thus, U.S. representation in Saudi
Arabia changed its role from mediation between the U.S. private sector and the local government to that of direct involvement in the economic affairs of that country, which
were now tied in with U.S. defense planning. The lend-lease
agreement with Ibn Saud, however, remained strictly within
the confines of international commerce. U.S. foreign policy
still continued to avoid any other than economic relations
with countries of this region. The Petroleum Reserves Corporation was founded under the aegis of this policy.

This breakaway from traditional U.S. foreign policy
goals was further punctuated by the controversy that it

caused at home. As in the case of the domestic Petroleum Code, this new controversy regarding government control of oil resources abroad was resolved in very much the same way: The private sector succeeded in excluding government intervention in its economic affairs. The same basic principles of the free enterprise economy also, basically, motivated Congress in its opposition to the proposed government ownership of such enterprises abroad.

Thus, U.S. foreign policy had to adapt itself once again to the proposition that any oil interests in the Middle East could be acquired either through the companies themselves or in opposition to them. Public opinion and petroleum lobbyists made it clear that only the first alternative was realistically feasible. In the first round, then, where the U.S. government sought to acquire managerial and proprietary rights over Middle East oil, the public sector lost to the private sector. In the second round, which was soon to follow, rather than negotiate again with uncooperative private concerns, the government elevated the question of oil to the diplomatic level. Here, at the negotiations table with representatives of another government, its official delegates spoke for the national interest in foreign oil reserves.

# 3

## IN SEARCH OF A U.S.
## FOREIGN OIL POLICY

The general disappointment in official circles generated by the failure of the Petroleum Reserves Corporation
to bring forth direct U.S. government influence on Middle
East oil affairs resulted in two simultaneous schools of
thought in Washington.  On one hand there were those who
thought that the government lacked a clear foreign petroleum
policy and that therefore it was prone to adopt schemes that
appeared to promise immediate high-ratio returns but could
not, in reality, be evaluated against any formulated set of
national policy principles.  It was now felt that such a
comprehensive long-range national petroleum policy was ur-
gently needed in order to avoid further failures and to
give sound direction to future decision-making.  Legislature committees, government offices, public agencies, and
national organizations became involved in a series of broad
investigations of every possible aspect of petroleum operations.  A nationwide effort was being made to prepare an
answer to the question of what the purposes of U.S. petroleum policy were in view of future domestic demands and expected postwar world shortages.
At the same time, while the government, the Congress,
and the public were still engaged in a major search to formulate such general principles of a national petroleum policy, the Department of State, spurred by the driving force
of Secretary of the Interior Ickes, began to seek, through
its diplomatic channels, an international negotiated agreement in order to determine the immediate and most pressing
postwar problems that were then expected to be involved in
the distribution of Middle East petroleum resources.
This chapter first explores the political background
and foreign policy objectives of the Anglo-American oil

pact, as were delineated by the political situation in the Middle East in general and in Iran in particular. Then, it looks into various attempts made in the executive branch, in the Congress, and by the private sector to formulate general guiding principles for a U.S. foreign oil policy. In the third section, this chapter deals with an abortive proposal made by the International Cooperative Alliance at the United Nations. The following section concerns the 1948 agreement signed by the international oil companies and its immediate consequences. The final two sections of this chapter first analyze the basic principles of U.S. foreign oil policy as it emerged in 1949 and then recount the nature of the Voluntary Petroleum Agreements that became the operative framework for all future cooperation between government and the companies under this policy.


CLOUDS OVER IRAN

Long before the end of wartime hostilities, U.S. and British oil companies began to seek new concessions in the promising areas of southeast Iran, especially in the province of Baluchistan.[1] The British Shell Company sent a negotiator to Teheran in the fall of 1943, and, in the spring of 1944, two U.S. oil companies, the Standard Vacuum Oil Company and the Sinclair Oil Company, followed suit with their team of representatives. At the same time, the Iranian government, in order to ascertain better its own natural oil resources, hired U.S. geologists A. A. Curtice and Herbert Hoover, Jr., "to survey the oil resources in various parts of the country."[2]

The news of the negotiations between the Iranian government and the British and U.S. oil companies was made public for the first time when Iranian Prime Minister Muhammed Said Marghai announced it to the Majlis (the house of representatives in the Iranian parliament) in August 1944. At that time, communist members of parliament in Iran opposed the granting of oil concessions to any foreign power. However, the Iranian Communist Party changed its position when, in the middle of September, the Soviet Union began to press for oil concessions in the provinces of northern Iran,[3] including the autonomous "democratic government" of Azerbaijan.

The first inkling that the Soviet application for Iranian oil concessions was postponed by the Iranian government until after the end of the war came from the Soviet press.[4] Soon after, on October 16, the Iranian government announced that "the granting of any concession would be held in

34

abeyance until after the close of the war."[5] The Iranian
parliament thereupon "approved a bill prohibiting any Iranian
official from negotiating or signing any agreement concerning
oil concessions under penalty of imprisonment," which would
extend to a period from three to eight years and would also
include deprivation of all state and official positions.[6] To
all three applications for oil concessions, the British, the
U.S., and the Russian, "the Iranian Government replied that
it could not consider requests until six months after the war
when, under the Anglo-Iranian agreement, all foreign troops
were supposed to be withdrawn from Iran."[7] (On January 29,
1942, Great Britain, the USSR, and Iran had concluded a
treaty of alliance whereby the first two powers had undertaken to respect Iran's territorial integrity and sovereignty and to evacuate their forces within six months of the
end of hostilities with Germany and its allies.)

The temporary Iranian rejection of all new oil concessions was followed by a bitter attack in the Soviet press,
asserting that the Soviet Union controlled only 11 percent
of the world's resources in petroleum[8] and accusing the Iranian government of being "pro-fascist" and "anti-Soviet."[9]
The Soviet journal War and the Working Class said that "the
law forbidding oil negotiations with any foreign powers, and
passed with 'unprecedented haste,' showed that statements by
Iranian authorities that they merely wanted to postpone signing concessions until after the war were 'hypocritical and
unfair' and that the true aim was to protect 'the monopoly
holding active concessions.'"[10]

The Soviet press responded bitterly to the Iranian government's postponement of new oil concessions until after
the war, and the overthrow of the government of Prime Minister Muhammed Said Marghai has been attributed to Soviet conspiracy. Indeed, according to the New York Times, the left-
wing party Tudeh, which greatly resented Said's attitude toward Soviet requests for oil concessions, "was largely responsible for the overthrow of the government."[11] However,
the new premier, backed by the British and U.S. governments,
also refused to bow to the Russian demands. The blame for
the unrest in northern Iran was laid directly at the door of
the Soviet Union. On November 19, 1945, Iran's new ambassador to Washington, Hussein Ala, made his first official
visit to the State Department to confer with Secretary of
State James F. Byrnes, at which time he made a charge that
"the revolt in the Iranian province of Azerbaijan had 'most
certainly' been 'engineered' by the Soviet Union."[12] Azerbaijan had been controlled by the Red Army since 1941, ostensibly in order to protect supplies sent from the Persian
Gulf to Russia by the United States through that area.

The news of the Soviet demand for oil concessions in
northern Iran made considerable waves in the U.S. and the
British press, as it was seen to constitute a major departure
from traditional Soviet foreign policy toward foreign concessions in general and toward concessions in Iran in particular.
Under Lenin, foreign capital had been permitted to obtain
concessions inside the Soviet Union. These concessions had
gradually been liquidated, ending with the Japanese coal concessions on North Sakhalin Island. No foreign concessions
had been requested by the Soviet Union. A major Soviet goal
in those days was to create a system of economic autarky independent of hostile external influences. To this end, the
Soviet government devoted all its energies to the development
of the internal resources of the Soviet Union and did not invest its funds outside the country. Moreover, Lenin also
gave up those concessions in northern Iran that had been obtained by the Czars. These were returned, presumably, on the
understanding that they were not to be handed over to either
governments or citizens of a third power.

Thus, the current Soviet demand for oil concessions in
Northern Iran reversed two of its traditional policies: its
withdrawal from Middle East concessions as well as its policy
of autarky. That the Soviet government would select oil in
order to effect this policy reversal, while Russia was more
self-sufficient in regard to oil than any other commodity,
caused a measure of consternation in both Great Britain and
the United States. The New York Times raised the question of
Soviet economic intentions in the Middle East: "Does Russia
wish new oil sources in order to have exports with which she
can pay for reconstruction goods? Does she intend to intervene politically in the Middle East and wish to establish an
economic basis for her intervention?"[13]

To Great Britain, however, the political goals of the
Soviet Union in northern Iran were at least as important as
the economic ones. For years Iran had been recognized by the
great powers, to the exclusion of Russia, as a British sphere
of influence. The Czars considered at least northern Iran as
being an area of their sphere of influence and challenged
Great Britain's rights over the rest of Iran. The 1907 pact
between Russia and England, among other things, also had allotted each signatory power its own zone of influence in Iran.
Yet Russia and England remained, until 1917, the principal
contenders for supremacy over Iran, Afghanistan, and Tibet.

Naturally, the Soviet demands for concessions in northern Iran evoked old political fears in Great Britain that
Russia was once again intent on competing with the Western
powers for supremacy of influence in Iran. Thus, the British

press viewed suspiciously Soviet religious liberalization
policy toward its minorities, including the Moslems, along
with its government's demands to reassert its protection over
orthodox monasteries in the Middle East.[14]  The British press
also noted with equal measure of alarm the promptness with
which the Soviet government had recognized the new independent governments of Syria and Lebanon.  In the British view,
the most important question arising from Soviet activities in
the Middle East was if this new policy was aimed at any particular political or economic objectives in that region or if
it was merely a means employed by the Soviet government to
pressure the Allied powers in other parts of the world.  On
this question, the Economist of London had this to say:

> There was always a sense in which, in the days
> of the Czar, the Russian policy in the Middle
> East was derivative.  They exercised pressure
> on the dominant Power in the Middle East, Brit-
> ain, in order to influence British policy in
> the vital European arena. . . .  Today, . . .
> Britain and, to some extent, America are the
> dominant Powers in the Middle East.  The better
> their relations with Russia in Europe, the less
> likely is Russia to fear for its security in
> the Middle East.[15]

This view of Soviet intentions in the Middle East was dis-
proved during the 1960s, when East-West tensions were progressively decreasing in Europe but were rapidly gathering
momentum in the Middle East.

Whatever Soviet intentions were in regard to Iranian oil,
the Soviet Union was severely criticized in the U.S. press
for the methods it employed.  The New York Times opined that
Russia's "policy in Persia has been traditional oil diplomacy
--at its best or worst.  Neither in seeking oil concessions,
nor in the methods it has used to force a reluctant country
to grant them, has it done anything which its western rivals
have not done in their time."[16]

To the New Republic, Soviet methods of claiming rights
to Iranian oil concessions "savored of the worst kind of economic imperialism."[17]


A NEGOTIATED INTERNATIONAL AGREEMENT

According to Feis, "All branches of the government now
became alive to the fact that any action taken in regard to

Middle East oil would command great political interest.
. . . Thus appreciation grew of the wisdom of seeking to attain our aims as part of a negotiated international agreement."18

The announcement by the government of Iran that all petroleum agreements with current and prospective concessionaries would be postponed until after the end of the war
aroused a great deal of anxiety in London and Washington.
Iranian oil supplied most of Britain's oil requirements,
while in the United States a new surge of oil panic, due to
a gloomy forecast of a rapid decrease in U.S. reserves,
caused grave concern about the availability of foreign resources. At that time Iran was the chief supplier of Middle
East crude petroleum and products. Accessibility to these
resources became a matter of a paramount national interest
to both the United Kingdom and the United States. The possibility that new-found oil resources in northern Iran might
be slipping away was a cause for alarm in both countries.
The presence of Soviet troops in northern Iran and the intentions of the Soviet government to acquire oil concessions in
that area seemed to cast a foreboding shadow. If petroleum
agreements were to be postponed until after the war, at a
time when the Allies were committed to pulling their forces
out of that country, it was feared both in London and in
Washington that the concessionary rights for oil exploration
and production in northern Iran would be ceded to the Soviet
Union along with all the political influence that these
rights implied. By the end of the war, the material pressures that the Allies would be able to exert over the Soviet
economy would be greatly lessened, while Soviet influence in
Iran steadily increased. England, then, for the first time,
welcomed U.S. intervention in Iran. Furthermore, the Shah
declared that Iran welcomed "signs that a positive and crys-
tallized United States foreign policy was developing for the
first time in the Middle East."19 Thus, World War II had
brought the United States into Middle East affairs not only
as a contender with Great Britain for Arabian oil but now
for the first time also as a welcome collaborator with England vis-à-vis the Soviet Union. This new alignment became
especially ostensible in regard to oil.

The strategic situation in the region was also becoming
conducive to cooperation. By the end of the summer of 1943,
the Mediterranean had been cleaned of enemy vessels, and, as
a result, there was a renewed U.S. interest in the possibilities for greater utilization of Middle East oil resources
by the United Nations for military purposes. Senator Ralph
Owen Brewster of Maine and Senator Henry Cabot Lodge of

Massachusetts, both Republicans, repeatedly demanded that, in view of the fact that U.S. oil reserves were getting low and because it was necessary "to conserve for future wars," greater use must be made of Middle East oil.[20] Again, a controversy ensued as to the proper way of acquiring those oil supplies for the national interest. And once again government, industry, and the press disagreed on methods and goals.

After the formidable obstacles that the oil companies had placed in the way of the Petroleum Reserves Corporation, it became apparent that no headway could be made in further negotiations between government and companies. The U.S. government, therefore, turned its attention to achieving its objectives through diplomatic channels on the intergovernmental level of negotiations. But agreements at the intergovernmental level could not be made unless those who gave their advice and consent and those who were most immediately financially concerned had first reached an understanding in regard to the proposed negotiations. President Roosevelt had learned the importance of reconciling the opposition from the bitter lesson of the Versailles Peace Conference. At that time, according to Secretary Byrnes,

> President Wilson neglected to invite the leaders of the political party in opposition to his administration to participate with him in making the peace.
> President Roosevelt, on the other hand, asked the congressional leaders to participate in the peace studies being made by the Department of State shortly after our entry into the war.[21]

President Roosevelt's pacifying tactics were adopted by many government departments of his administration. Just as the president sought to avoid opposition by encouraging Congress to endorse a bipartisan position on issues of foreign policy, so also his secretary of the Interior desired to prevent an opposition to the government's diplomatic efforts by attempting to consult and reconcile the oil industry in general and the petroleum lobby in Washington in particular. Cognizant of the obstacles that the petroleum industry was able to put in the way of the Petroleum Reserves Corporation, Ickes, in his capacity as Petroleum Administrator for War, appointed a foreign oil policy committee composed of the leading international petroleum industrialists of the country, to submit for government consideration its own

recommendations for a postwar international petroleum policy
for the United States.

The report of the committee, however, ran counter to
Ickes's theory "that the government, through the recently
formed Petroleum Reserves Corporation, should enter into oil
business, particularly in the foreign field."[22]  Therefore,
Ickes refused requests by the committee and other interested
parties for releasing the report for publication and, in
fact, requested that it be "restricted."*

What the 12-member foreign operations committee, headed
by Orville Harden, a vice president of Standard Oil Company
of New Jersey, had recommended was "a plan for the establishment of an international oil compact similar to the Interstate Oil Compact of 1935 for the 'efficient and orderly development of the world's oil resources.'"[23]  It suggested no
preferential treatment for U.S. nationals as compared with
the nationals of other countries and supported the principles
of competitive free enterprise.  The report also pointed out
that U.S. military power would be enhanced by having "adequate and strategically located sources of oil supplies
throughout the world" but insisted that these sources ought
to be placed in the hands of private U.S. companies free of
government intervention.  Of the three possible methods of
developing foreign oil resources--by government operations,
private enterprise, or mixed operations--the report emphasized that "private enterprise was the surest and soundest
choice."

> Private enterprise has developed the requisite
> managerial skill and operating knowledge, and
> is already well established.  It can operate
> with a minimum of political complications,
> as most foreign countries readily admit for-
> eign capital, but few countries, if any, would
> look with favor upon operations by alien gov-
> ernments.  The greatest immediate need on the
> part of private enterprise is the assurance
> of our Government that it will seek to reduce
> the political risks involved in the use of
> private capital abroad, and that it will take

---

*J. H. Carmical, oil correspondent for the New York
Times, who had gained access to the report of the foreign
operations committee, reported its findings and recommendations, thus making its contents public.

no steps to discourage the efforts of the
nationals of the United States to maintain
an effective and serviceable oil industry in
foreign countries.[24]

The report urged special cooperation with Great Britain
on the ground that the largest part of the known oil reserves involved in international trade was "owned, controlled or under concession" held by the nationals of the
United States and Great Britain. However, this special cooperation between the two powers, the report stated, ought
to be expanded into an international oil compact to be adhered to by all countries, producing and consuming alike,
for the "equitable distribution of oil to all nations and
the avoidance of national restrictions." The international
agreement would then be patterned along the model of the
Interstate Oil Compact of 1935, which

does not oust our States from control of their
own policies. Similarly, each country must be
left freedom with reference to its own policy.
Within the United States the oil compact as-
sumes the coordination of policies of the
various States, with Federal encouragement.
Similarly, the oil policies of various countries could be coordinated.[25]

In the meantime, the petroleum committee of the International Economic Affairs branch of the Department of State,
under the direction of Herbert Feis, reverted to what was,
in essence, the principles of the open-door policy, under
which the United States had first made its claims for stakes
in the Middle East. Repeating the same argument that had
been pressed by the U.S. government after World War I, to
the effect that the U.S. contribution to a common victory
entitled U.S. enterprises to an "equal economic opportunity"
in the Middle East, the United States became thus an economic interloper in Middle East political affairs. However,
it also realized that in view of the prevailing attitudes of
the nations concerned as well as public opinion at home, the
United States should, if possible, "make its way in the Middle
East as a welcome and reconciling partner, and not stumble
into it as an intruder."[26] There was a general judgment
throughout much of the department "that it would have been
wiser to seek a solution for our problem by a more cooperative method."[27] Therefore the State Department Petroleum Committee recommended that "an international basis

should be sought for an orderly and combined program. Since British companies controlled so large a part of the oil resources of that area and British political influence was so widely established therein, it seemed sensible first to seek agreement with Great Britain."[28]

Rumors, however, had it that previous meetings at the Cairo and Teheran conferences had already laid the groundwork for a U.S.-U.K.-USSR "program for the development of the vast oil resources of the Middle East."[29] The details of the program, reported the New York *Times*, were to be worked out "in a series of bilateral conferences along the lines of the pre-war trade treaties conducted by the State Department."[30] And, indeed, Acting Secretary of State Edward Riley Stettinius announced on February 11, 1944 that discussions with the British regarding the whole petroleum situation would soon begin and that the Russians were also expected to participate at a later stage.[31] What was involved in the U.S. idea of international oil negotiations among the governments of the United States, Great Britain, and the Soviet Union was an agreement regarding "the sale and distribution of petroleum products in the world markets."[32] According to the New York *Times*:

> With an understanding in principle of the
> over-all scope of such a plan reported to
> have been reached at the Teheran and Cairo
> conferences, . . . it is understood that the
> details will be worked out . . . in a series
> of bi-lateral conferences along the lines of
> the pre-war trade treaties conducted by the
> State Department. It is understood that the
> Russian mission is scheduled to follow imme-
> diately after the conclusion of the confer-
> ences with the British. . . .
>    When sounded out on what participation
> it wanted in the export markets, it is under-
> stood that the Soviet Union suggested that
> its share should be 20,000,000 tons annually,
> or about 150,000,000 barrels. This is at the
> rate of a little better than 400,000 barrels
> daily. In the pre-war period Russia's produc-
> tion amounted to only about 250,000,000 bar-
> rels yearly, or around 700,000 barrels daily.[33]

The same anticipation of a broad international agreement between the major nations involved in oil explorations that existed in Washington also prevailed at that time in

London.  According to the _Economist_, these agreements were
intended to include not only Great Britain, the Soviet Union,
and the United States but also eventually Holland and Mexico.[34]

In the meantime, in November 1943, the Department of
State invited the British government "to send delegates to
Washington to discuss problems of mutual interest in the
field of international petroleum."  This discussion fell
into the terms of Allied peace planning as expounded by principles of the Atlantic Charter, the fourth article of which
had announced the signatory powers' intention to endeavor,
"with due respect for their existing obligations, to further
the enjoyment by all States, great or small, victor or vanquished, of access, on equal terms, to the trade and the raw
materials of the world which are needed for their economic
prosperity."[35]  Not least among the issues that were intended at the time to be examined in this light was the international petroleum industry.  No specific mention was
made in the U.S. invitation to Middle East oil.  The British
government accepted the invitation, and in May 1944 negotiations were begun.

The ostensible aim of these negotiations, Professor
George Lenczowski observed, was

> to promote an orderly development of the inter-
> national oil trade, to assure equal opportunity
> in the quest of new concessions, and to ensure
> adequate supplies of oil to all countries.  The
> agreement did not mention the Middle East spe-
> cifically.  Nevertheless the real motive, so
> far as the United States Government was con-
> cerned, was to assure for American interests
> a fair share in Middle East oil exploration
> and to obtain from Britain--hitherto regarded
> as the paramount power in the area--recognition
> of the legal and political validity of existing
> American concessions especially in Saudi
> Arabia.[36]

In concrete terms, the details that were expected to be
discussed at these meetings were the construction of two
pipelines from Iran and Iraq to the refinery at Haifa and the
expansion of the Haifa refinery to a 350,000 barrels per day
capacity.  If the United States was indeed going to draw
larger amounts of petroleum and petroleum products from these
sources, as was persistently urged by Senators Brewster and
Lodge, it was necessary to arrive at an agreed arrangement

with Great Britain, which held both the commercial control over the supplies and the political influence over the region. But the draft that emerged from these conferences disappointed all those who hoped for practical answers to Middle East oil questions.

In Washington, the purpose of the first Anglo-American Petroleum Agreement of August 8, 1944,[37] was announced to be the prevention of "friction between nations growing out of the problem of foreign oil and assuring to all an adequate supply." The Introductory Article to the agreement stipulated that petroleum supplies "should be available in accordance with the principles of the Atlantic Charter and in order to serve the needs of collective security" (paragraph 4); the agreement further stipulated that "adequate supplies of petroleum shall be available in international trade to the nationals of all peaceloving countries at fair prices and on a non-discriminatory basis" (paragraph I, 1); that benefits received by the producing countries from their petroleum resources should be used "to encourage the sound economic advancement of those countries" (Article I, 2); and that "the principle of equal opportunity shall be respected by both Governments" (Article I, 4). Finally, the signers agreed to establish an International Petroleum Commission whose tasks would be to prepare long-term estimates of world demand and commensurate recommendations of ways and means to satisfy those demands "by production equitably distributed" (Article III, 1 and 2); and to analyze short-term problems of production, processing, transportation, and distribution of petroleum on a worldwide basis (Article III, 5 and 6). The agreement was signed for the United States by Stettinius, then acting secretary of State, and for the United Kingdom by Lord Beaverbrook, who was then Lord Privy Seal.

The agreement incorporated the principles of petroleum distribution on "an equitable and non-discriminatory basis" to "all peaceable countries." In reply to a question before the Senate Special Committee at its hearing, Charles Rayner, Petroleum Adviser to the Department of State, explained that this statement was essentially a reiteration of the open-door policy, while the reference to "equal opportunity" referred to "the opportunity of finding the oil, of going in and exploring for the oil."[38] On the other hand, critics of the agreement argued that by this clause Great Britain and the United States could not only apply various meanings to the phrase "all peaceable countries" but could also use their control of the world's largest petroleum resources in order to enforce their definition of who was peaceable and who was warlike.[39]

However, a theory was advanced along with the instrument that by an international agreement between the United States, Great Britain, and the Soviet Union, the control of all oil resources could be "so solidified that no nation would make a step toward disturbing the world peace."[40] Since it was estimated at that time that 95 percent of the then known reserves of the world were controlled by the British, the Americans, and the Russians, an oil agreement between the first two nations, the government spokesmen argued, constituted an important step toward the desired goal of a stable peace.

As soon as the agreement had been signed by the acting secretary of State, the Department of State envisioned the ensuing enlargement of its scope. The Department's economic adviser, Herbert Feis, recommended that "the full participation of the USSR, France, and the producing oil countries should be solicited."[41]

The British press was in full accord with the concept of an enlarged international base to the oil agreement. It had fully expected the participation of official representatives from Russia, the Netherlands, and Mexico at the Washington Conference.[42] However, when the negotiations were narrowed down to an Anglo-American agreement, the British press commented that it was "obviously desirable that countries such as Russia and the Netherlands, which are intimately concerned, should be drawn into the discussion at the earliest opportunity."[43]

Viewed critically at the Kremlin, the suggestion that the Soviet Union would be serving the cause of world peace by affixing its signature to the Anglo-American Oil Agreement might very well have caused a great deal of skepticism. Soviet summation of the proposal extracted it from the vague language of the Atlantic Charter and placed it in the midst of a concrete Middle East situation: Whatever the meritorious effects of an international petroleum agreement between all these countries might have been for the future stability of world peace, there was no question that once adhered to by the Soviet Union, it would have denied the Russians exclusive rights to any concessionary options over northern Iran's petroleum. It would be at that point that the United States, the United Kingdom, and France would have stepped in and, aided by principles of the Atlantic Charter that underlay the agreement in regard to the equitable distribution of world resources, would have claimed that the Soviet Union, having more petroleum than any one of them and many more times than it needed for its low level of consumption, should in effect share its concessions with those countries that had greater need for those resources.

The idea that a Western initiative could draw the Soviet Union into an oil agreement with the United States and the United Kingdom, the one viewed by the Kremlin as the most capitalistic and the other as the most imperialistic nation in the world, signified some basic lack of understanding of Soviet thinking prevailing in Washington in those days, which was also mixed with hopes for peace and cooperation within the framework of a worldwide international organization. In fact, two views shaped Soviet policy toward the Western nations. On one hand, there was a deep-rooted conviction that the Soviet Union was an isolated socialist island surrounded by a hostile sea. At any opportunity, those menacing hostile waves would rise to engulf the island. The Soviet Union could not, therefore, drop its guard. It could not trust the professed good will of any of these fundamentally hostile regimes. To let any part of its economy be tied up with Western markets and their price fluctuations would set that part of the Soviet economy at the mercy of these powers. It would be suicidal to that sector and might have an adverse impact on the whole Soviet economy. According to this view, it would be naïve to expect that the Western governments, once given the power attained by an integrated trade agreement, would refrain from exerting pressure on the Soviet Union in an attempt to strangle the entire Soviet system.

The other view that shaped Soviet policy was the Marxist tenet that wars between capitalist nations would precipitate their downfall. Stalin fully expected World War II to have set in motion forces that would bring about the collapse of the capitalist regimes.

Given these premises, it would have been a foolish step for the Soviet Union to get itself involved in trade agreements with governments who were not only hostile to it by definition but also whose days were numbered.

As soon as the Anglo-American Agreement had been published, it stirred such a violent public reaction throughout the country that the Department of State considered it prudent to yield to the demand of Senator Tom Connally of Texas to submit it for ratification by the Senate as a treaty, rather than have it finalized as an executive agreement. On August 24, 1944, the president therefore submitted the agreement to the Senate for ratification, where it immediately encountered a storm of angry opposition from the petroleum lobby.* The

---

*According to the New York _Times_, December 11, 1944, the petroleum lobby's objections were most systematically set out in a resolution of the Petroleum Industry War Council, which called upon Congress not to ratify the treaty.

executive branch and the oil companies did not seem to see eye to eye on the instrumentalities of control of foreign oil implied in the agreement. While Washington sought a program that would mean "greater direct participation of the government in the foreign oil picture," the industry wanted the appointment of an "international petroleum commission with advisory powers only."[44] According to Feis,

> The industry felt that if the government ob-
> tained extended powers of control, these would
> probably be used. It wished for vigorous and
> unfailing government support in foreign opera-
> tions, but without subjecting itself to the
> will of the government.
> There was ground for the belief that the
> operation of the agreement might be used to
> curb the future freedom of private oil inter-
> ests in some respects.[45]

Moreover, the petroleum agreement had been drafted with virtually no advice from the U.S. oil industry and with a total disregard for the findings and recommendations of the Committee of Petroleum Industrialists (see above page 39) formed by Secretary Ickes. The agreement succeeded in re- newing the oil industry's suspicions of another Rooseveltian attack upon its freedom, a renewed attempt to subordinate it to government controls, which it was determined to resist in time of peace. It also totally disapproved of the composi- tion of the U.S. delegation to the petroleum negotiations.

These negotiations, between Great Britain and the United States, had opened on May 3, 1944; the representatives of the two nations were not of equal standing either in government or industrial experience. While the British sent to Washing- ton two of their most experienced oil men, Sir William Fraser, chairman of the Anglo-Iranian Oil Company, and Sir Frederick Godber, managing director of the Royal Dutch-Shell Oil Com- pany, the U.S. delegation consisted of government officials and military men who had limited and largely only administra- tive experience with oil.[46] It appeared that it was the U.S. government intention to exclude representatives of the inter- national petroleum industry from the conference and to con- fine the negotiations to considerations of international politics and security.* On the other hand, according to

---

*Attorney General Francis Biddle, viewing the question of participation by oil industrialists and State Department representatives in policy-making formulation on the

Michael Brooks, the tendency in England appeared to be "to regard oil as a technical and economic problem . . . but not as an integral part of a politico-strategic picture. . . ."[47] The ensuing agreement succeeded remarkably in displeasing nearly everyone who was associated with the oil industry on both sides of the Atlantic. In the United States, there was a sharp criticism of its ambiguous and vague language, while in England there was a feeling that "if the art of diplomacy consists in disguising intentions, then the Anglo-American oil agreement is a masterly document."[48] Obviously, the U.S. government was unable to sign an oil agreement without the prior approval of it by the industry most involved in it. In order to arrive at such an international agreement, then, it was first necessary to overcome the opposition of the oil industry. To that end, Ickes spent the following months in discussions with representatives of the industry.[49]

But it was not the oil industry alone that had to be won over. Public reaction to this agreement was as violent as in the case of the proposed pipeline to the Mediterranean. The protests raised by the industry and the public against the agreement were forceful enough to impel the two governments to initiate new negotiations. On January 10, 1945, Roosevelt asked the Senate to return the agreement, and a revised agreement was worked out informally between the Senate Foreign Relations Committee, representatives of the major international oil companies, and the Department of State.

In September of that year, Ickes, heading the U.S. delegation, left for London to open the renegotiations of the Anglo-American Oil Agreement. The revised agreement was drafted on both sides with the advice of the representatives of the international petroleum industry, who eliminated from its text all references to the Atlantic Charter and its principles and put the new instrument on a purely economic basis, which reinforced the freedom of private capital.[50] One of the first acts of the Labour government after it came to power in England was to sign the new agreement.[51]

---

intergovernmental level, warned against violation of the antitrust laws. The opinion taken by the Department of Justice was that "the formulation of policy decisions remains with the appropriate government agency," that "authority should be vested exclusively in public officers who are responsible to the President and to Congress," and that any consultation with the industry would be properly made only by means of an Advisory Committee representing private interests but having "no authority to determine national policy or to carry out a program." (New York Times, April 27, 1944, pp. 26 and 27.)

But in the United States the revised agreement caused
a new wave of criticism.  If the first agreement had succeeded in displeasing those connected with the international
petroleum industry of the Middle East, the second agreement
displeased the smaller "independent" companies, which feared
government favoritism toward the foreign petroleum companies
and, at the same time, suspected a hidden tendency in the
agreement to establish restrictive practices that would narrow their own opportunities for expansion into new avenues
of international trade.  They suspected the agreement would
lead to a large expansion of Middle East oil production, as
a result of which "cheap" Arab oil would flood the domestic
markets.[52]

President Harry S. Truman sent the new agreement to the
Senate Committee on Foreign Relations in November 1945.  The
committee did not seem able to include it on its agenda until
June 1947.  After nearly two years of delays, a new proces-
sion of witnesses, and a smothering consideration in long
hearings, the committee at last approved the agreement and,
after having made some changes in it, returned it to the
Senate floor, where it remained forever pigeonholed.


## AN OIL POLICY FOR THE UNITED STATES

The Anglo-American Petroleum Agreement found its burial
grounds in the Senate, but the question as to who should
rightly chart the petroleum policy of the nation--the oil
men, who were experts in the technical field and were versed
in its economics, or the federal government, which was concerned with the international situation of the country--
started a lively debate in the press.  The foreign operations
committee of oil industrialists supported the view that the
question of oil was essentially a commercial problem and
should, therefore, be left in the hands of the producing
companies.  Thus, its report listed five guiding principles
for an immediate foreign oil policy for the United States:

> 1.  The American petroleum industry
> should be encouraged to expand its plans for
> developing the world's oil resources.  This
> encouragement requires assurances that na-
> tionals of the United States will receive the
> cooperation of our Government in securing a
> position of equal opportunity with the nationals
> of other countries and that the Government it-
> self will not enter into competition with its
> nationals.

2.  Existing handicaps in the oil operations of our nationals abroad should be examined in the light of their incidence upon our
national welfare and efforts should be made to
remove such handicaps as originate in the laws
and practices of the United States.  American
nationals operating abroad must be able to comply with the laws and customs of foreign countries without incurring the risk of violating
American laws.
3.  The diplomatic support accorded to
our nationals by the Government of the United
States in the course of the war should be as
effective as that accorded to nationals of
other countries by their respective Governments.
4.  The ultimate disposition of oil facilities and supplies developed or paid for abroad
by the Government of the United States in the
course of the war should be so designed as to
promote the interests of our nationals in the
post-war world.
5.  Our foreign oil policy should also
include appropriate measures for the return
after hostilities of American-owned properties,
rights and interests in Axis-held territory;
adequate compensation for assets destroyed and
damaged; and assurance of an early commercial
operation of returned properties.[53]

These five guideline principles were exclusively concerned with the private interests of the oil companies.
None of them dealt with the national interest or questions
such as forecasts of future domestic supply and demand or
procedures in the event of a new national emergency.

In the prevailing controversy, Bernard Brodie held the
opposite view.  He suggested that "strategic considerations
must command first attention. . . .  economic interest involved a consideration mainly of price differential . . .
while the strategic interest concerns the life or death of
the nation."[54]

In evaluating the strategic properties of foreign petroleum, however, he raised the question as to which petroleum must be protected in time of war and at what cost.  The
dilemma of the dependence of national defense on foreign
petroleum would involve the United States in the primary
concern of protecting "sea-borne communications from the
Middle East to our shores--a line which except for our

dependence on oil we might have no military interest in maintaining."[55]   Therefore, in his opinion,

> The strategic approach to the oil problem must
> be based on the premise that, so long as it
> can be made to fulfill our basic wartime needs,
> the only oil reserve worth defending is that
> which can be held with a minimum of defensive
> military commitments.  That portion of it
> which falls within the area we must in any
> case defend is pure windfall strategically.[56]*

In the Department of State, Herbert Feis, however, attached far greater significance than Brodie to Middle East oil, both in respect to the conservation of domestic reserves and resources and in respect to times of emergency. "The proper guiding aim of policy," he said, "is that this country and its armed forces should, during a time of emergency, have ample supplies, wherever needed, with the least military risk or burden."[57]

> But in determining the usefulness of foreign
> oil supplies to us <u>during</u> the period of ac-
> tual emergency, their <u>location</u> is obviously
> all-important.  Only sources of supply within
> our assured military control and accessible
> to our armed forces would be of use.
>       Each different military combination
> and campaign would pose its own problem.[58]
>
> The Middle East is potentially important to
> us as a source of oil supply that supplements
> our own and prolongs our reserves.  The war
> has established the fact that American mili-
> tary action may take place anywhere in the
> world, and that, particularly in any struggle
> involving the Pacific, control over these

---

*A U.S. line of communication to the Middle East had existed before oil became important.  However, its chief purpose was to protect commercial U.S. interests in that region. Since such interests were small, the economic significance of the line was also small.  Moreover, since no strategic commodity was involved in this commerce, that line of communication was not vital for the national defense and could be easily abandoned without much loss to U.S. interests.

oil-fields (and the political status of this area) might be of direct concern to us.[59]

In the absence of guiding principles for a comprehensive foreign oil policy by the United States, the Department of State tentatively operated upon the assumption that since the oil reserves of the United States were, presumably, insufficient to meet future growing demands, it was in the nation's best interest to look for oil fields outside the Western Hemisphere. Of these fields, Saudi Arabia presented the best potential. Once this premise had been accepted by the Department of State, this area became important to the national interest.

The Senate Special Committee Investigating Petroleum Resources, one of the many groups occupied with the problem of formulating a national petroleum policy, did not deny the State Department's premise concerning the uncertain supply of domestic petroleum resources. But in accepting it as a working premise for future policy-making, the committee came to assume that the main problem was a military one, and it suggested going beyond the national defense to keep "aggressor nations" from getting control of the oil.[60] Reliance on foreign oil to any sizable extent could not be accomplished successfully without a total reversal of the U.S. defense posture. Up until that time, this country had considered the Pacific as its area of vital concern. But in moving into the Middle East, where the greatest oil reserves were located, the whole military problem was "radically transformed, in view both of the distance and of the character of the terrain."[61] The investigating committee feared that this military transformation might lead to the trebling of the military establishment in order to meet the "necessity of a military defense of the Arabian area"[62] and to relying on long tanker lines and "the maintenance of a tremendous fleet."[63]

However, the War Department denied that the shift in oil strategy implied an increase in military commitments. It supported the navy's view that it was highly desirable to have oil resources at its disposal scattered around the world for times of emergency, "making it unnecessary to have that long haul from the United States to those areas."[64] Those foreign resources, it was argued, would secure for the United States, in case of future war, the availability of sufficient petroleum for all national defense purposes, including civilian as well as military."[65] Not only was the War Department in favor of continuing such foreign operations, but it also was firm in its assertion that the acquisition of these foreign reserves would not cause any increments in military commitments in that area. While asserting that if "we should lose oil reserves in the Middle East, presumably they would then belong to someone else . . . to our great disadvantage,"[66] it also reassured the Senate special committee that it was not in any way contemplating having bases there.

This political view was supported by the realities prevailing in the Middle East at that time, when England and France still ruled over the area and were assumed to be fully capable of maintaining its defenses. For, historically speaking, U.S. military commitments in these countries came about years later as a result of the extension of the policy of containment from the Eastern Mediterranean region to the entire Middle East, whereupon it "grew slowly and fitfully from crisis to crisis." When gradually British and French bases began to disappear, particularly after 1956, only then the "balance of commitment shifted from Britain and France to the United States."[67]

In its <u>Final Report</u>, the Senate's Special Committee Investigating Petroleum Resources took a somewhat similar view of the strategic use of foreign oil to that which had been expounded by the petroleum liaison officer for the War Department. In the committee's opinion,

> If the United States should become engaged in
> a war waged wholly or partially outside its
> boundaries, the availability of oil from Amer-
> ican reserves near the foreign theatre of
> hostilities would be advantageous. To the
> extent that the war were engaged within or
> near the continental United States, the re-
> serves within this country would be of a para-
> mount importance.[68]

While thus seeming to support the War Department's view of the strategic properties of foreign oil, the committee, in fact, initiated a new departure from recent trends of opinion inasmuch as it ceased to rally around the cry about the scarcity of domestic resources. It ignored the prevailing panic about world oil shortages and instead placed its main reliance on encouraging the development of domestic rather than foreign petroleum operations. The committee recommended that in order to guarantee "a domestic petroleum supply adequate for all eventualities" the following steps must be taken:

> (a) Incentives to promote the search for
> new deposits of petroleum within the boundaries

of the United States and in the continental
shelf; and
(b) The continuation of the present
program looking to the manufacture of synthetic liquid fuels to supplement our domestic crude supply.[69]

In order to facilitate this program, the committee proposed the following:

(a) New exploratory drilling on the public domain and elsewhere;
(b) Deeper drilling, as a result of improved technology;
(c) Stimulated production from old fields
by improved methods of secondary recovery and
by payment of governmental subsidies;
(d) Exploration of the continental shelf;
(e) Manufacture of gasoline and other
products from natural gas; and
(f) Extraction of petroleum from oil
shale, and manufacture of synthetic liquid
fuel from coal as well as from agricultural
commodities.[70]

While shifting the emphasis of U.S. oil policy from
the Middle East back to the domestic resources of the country, the Senate special committee was thus, in fact, minimizing the value of foreign resources for the national interest.
According to this view, the protection of these foreign reserves did not occupy the same place of high priority as was
indicated in the testimonies presented by the Departments of
State, the Interior, War, and the Navy. Statements made by
representatives of these government departments before the
special committee had reiterated the importance of foreign
oil to national security and the general welfare of the nation. They asserted that heavier reliance on foreign resources would also entail government control of those private
oil enterprises. They manifested no hesitation in recommend-
ing that protection of the interests of U.S. oil companies
operating abroad should be made conditional on their close
cooperation with government policies. In fact, the War Department even went a step further in suggesting that "as a
corollary to governmental protection, all possible arrangements should be made to prevent American nationals from dis-
posing of their foreign oil resources by sale or lease without the approval of the United States Government."[71]

But the special committee made no such recommendations in its <u>Final Report</u>. As it was inclined to limit U.S. national interest in foreign oil resources, it felt at greater liberty to emphasize the right of these private oil companies operating abroad to be free from government intervention. It even rejected the notion that government intervention in the trade practices of the international petroleum industry was warranted on grounds of alleged violations of the Sherman Antitrust Act. "Size alone," the committee reported, "does not constitute monopoly."

> It is the unfair methods of trade, which seek
> to destroy or exclude competitors by means of
> intercorporate stockholdings, or by agreements
> between present and potential competitors,
> whereby control of commerce among the states
> or with foreign countries is secured, that are
> anathema to the people of the United States.
> But how far shall we go in advancing beyond
> our borders the principles of laws against re-
> straint of trade?[72]

The problems delineated by the Senate special committee in respect to alleged monopolistic practices were, in essence, two separate ones. The first one raised the question of what, in fact, did (or did not) constitute a monopoly in violation of the Sherman Antitrust Act. The second question that it raised concerned the extent of power of the national jurisdiction to deal with such violations by U.S. companies beyond the national territory. The committee's restrained position reflected the large oil interests represented in the Senate. In its moderation it stood in sharp contrast to the judicial opinion expounded by the Supreme Court. In the case of U.S. vs. Socony-Vacuum, both the definitions of acts violating the Sherman Antitrust Act and the extent of power of the United States to deal with such acts as committed outside the U.S. boundaries received clear answers. (For the Court's opinion in this matter see this chapter, "An International Petroleum Industry Agreement" below.)

The Senate Special Committee Investigating Petroleum Resources had begun operations on March 13, 1944. Nine reports and nearly three years later, on January 31, 1947, the committee concluded its work. Its most important recommendation was that a national petroleum policy, based on a reorientation toward domestic resources, was necessary.

AN INTERNATIONAL SOLUTION:
A UNITED NATIONS INTERLUDE

In his Thirteenth Report to Congress on Lend Lease
since the program had been initiated in March 1941, President Roosevelt foresaw an international agreement that would
give all nations equal access to the world's oil supply.

> Agreed action by the nations of the world, as
> provided for in the master lend-lease agree-
> ments, for the expansion of production, the
> elimination of discriminatory treatment in
> commerce and the reduction of trade barriers,
> will assure to the United States and other
> nations fair and equal access to the petro-
> leum produced in all parts of the world.[73]

Oil companies' executives declined to comment at the
time on President Roosevelt's plan for pooling the world's
petroleum resources among all nations, producing and consuming alike, in accordance with principles already expounded by the Atlantic Charter. Subsequently, however,
other references to similar arrangements of world distribution of raw materials were made intermittently both in the
press and in official circles. The suggested plans ranged
from bilateral agreements between designated governments to
an international commission appointed by the United Nations
to supervise all the world's natural resources.
    Much of this talk had disappeared shortly after the end
of armed hostilities and the beginning of the Cold War. But
some vague talk lingered on for some time, both in the press
and in official circles in Washington, to the effect that
the question of petroleum resources, along with the question
of all world raw materials, strategic and otherwise, ought
to be handed over to the United Nations for equitable distribution among all people. However, an egalitarian policy
of distribution of all raw materials in the world to all
people according to need not only would have meant a share
in the wealth of African and Asian nations but would also
have signaled a U.S. commitment to include the natural resources of the United States in the world pool. Neither
the mining industry of the United States nor the U.S. government was prepared to take this step. The initiative in
this direction was taken in respect to petroleum alone by
an organization that had no resources of its own to lose.
Item Twenty on the provisional agenda of the United Nations
Economic and Social Council Fifth Session called for a

discussion of a proposal submitted to it by the International
Cooperative Alliance* to create a United Nations Agency, or
an "Authority, responsible to the Economic and Social Council,
for administration of the oil resources of the world, to begin with the oil resources of the Middle East, by and with
the consent of states involved."[74] The estimated Middle
East petroleum reserves in 1945 are given in Table 2. The
International Cooperative Alliance urged the Economic and
Social Council of the United Nations to make it possible for
cooperative organizations to secure practical means of giving
the pledge of the Atlantic Declaration in order to secure the
prerequisite economic conditions for an "enduring peace between nations."[75] Raw materials, the alliance stated,
"should be the first thing after armaments to be placed under
the control of the United Nations."[76]

TABLE 2

Estimated Middle East Petroleum Reserves, 1945
(millions of barrels)

| Company | Amount |
|---|---|
| Anglo-Iranian Oil Company | 27,750 |
| Royal Dutch-Shell Group | 2,750 |
| Gulf Oil Corporation | 5,000 |
| Standard Oil of California and Texas Company | 20,500 |
| Standard Oil Company of New Jersey and | |
|   Socony Vacuum Oil Company | 2,750 |
| French Companies* | 2,750 |
|   Total | 61,500 |

*Since Standard Oil Company of New Jersey, Socony Vac-
uum Oil Company, and Royal Dutch-Shell held stock in some of
the French concerns, there were indirect U.S., British, and
Dutch interests involved in the oil held by the French.

Source: J. H. Carmical, The Anglo-American Petroleum
Pact (New York: American Enterprise Association, 1945),
p. 46.

---

*Rule 10 of the rules of procedures of the Economic and
Social Council grants to permanent consultants the right to
present items for the agenda.

"In the land mass connecting Europe and Asia is a power prize second only to atomic energy itself--proved crude oil reserves that run to some 26 billion barrels, with potential reserves that may top 150 billion barrels."[77]

Further, the International Cooperative Alliance expressed the opinion "that the exploitation of the world oil resources might have been far more successful had it been undertaken in the spirit of planned international collaboration, instead of the spirit of battles, conquests and defeats."[78]

By 1946-47, the situation in northern Iran was completely altered. The political conditions that, in 1943 and 1944, had prompted Great Britain and the United States to seek a solution by means of a contrived Anglo-American Oil Agreement no longer existed. The Allies' anxiety over the oil concessions and the general security in Iran had subsided considerably. The Anglo-Iranian Oil Company was again the chief producer of crude oil in Iran and supplied most of the Middle East petroleum products from its refineries in Abadan, while the social unrest that had threatened to abolish the concessions had been temporarily curbed. At the same time, the U.S. government was exerting pressures on the Soviet Union to abide by its wartime agreement to withdraw all foreign troops from Iran upon termination of armed hostilities.

To the Soviet Union, however, the situation in Iran seemed increasingly unsatisfactory. Its influence there was rapidly diminishing while international tensions had now gained a threatening momentum under the impact of the Cold War. The need to fortify its borders became the paramount concern of Soviet foreign policy.* But if the Soviet Union

---

*That the security requirements of the Soviet Union were paramount in the minds of Soviet leaders in the postwar era was intimated by Soviet Foreign Minister Vyacheslav Molotov to American Secretary of State James Byrnes at the London meeting of the Council of Foreign Ministers. On December 24 Stalin expressed to Byrnes a similar disquietude about Soviet security. (See James Francis Byrnes, Paris Meeting of Foreign Ministers, a report, Department of State Publication 1537, Conference Series 86 [Washington, D.C.: Government Printing Office, 1946], p. 10.)

Constant Soviet concern about the security of its borders and the integrity of its territory prompted Byrnes to state publicly that "we deplore the talk of the encirclement of the Soviet Union." (James Francis Byrnes, Paris

had been successful in erecting a contiguous chain of defense along its western periphery, it completely failed to form a similar belt of friendly buffer states on its southern flank. The continued presence of large British and U.S. petroleum concerns in Iran and Iraq, coupled with the imminent withdrawal of its own forces from the Middle East, the civil war in Greece, and the recently proclaimed Truman Doctrine, emphasized the dangerous vulnerability of the Soviet southern exposure. Having failed to change the regime in Iran, the Soviet Union's next step was to attempt to neutralize the power the oil interests had in that country. That was done by broaching to the United Nations, through the International Cooperative Alliance, a proposal that, if accepted, would have decreased the power of the foreign interests in Iran and, at the same time, brought the Soviet Union back into Iran under the umbrella of international administration. By agreeing to pool all Middle East oil resources together for equitable distribution according to need, the private international petroleum industry would have lost its traditional role in the economy and the Great Powers would have assumed the managerial responsibility for the gigantic enterprise. The proposal of the International Cooperative Alliance failed to evoke enthusiasm in the United Nations, where the Western powers held the majority of votes.

Very little else was heard of this idea inside the UN walls.* The question of a world pool of natural resources was allowed to sink in committee. The intensification of the Cold War, on the one hand, and a series of petroleum crises in Europe and the Middle East, on the other, brought all hope for an international collaboration on oil policies to an abrupt end.

---

Peace Conference, a report, Department of State Publication 2682, Conference Series 90 [Washington, D.C.: Government Printing Office, 1946], p. 10.)

*The International Labor Organization's Committee on Petroleum Production and Refining was occupied with questions of labor-management relations in the petroleum industry and attended to requests from governments for technical assistance to utilize nonagricultural resources, including petroleum and coal reserves, as well as the training of personnel for surveys undertaken for this end.

## AN INTERNATIONAL PETROLEUM INDUSTRY AGREEMENT

In the meantime, because of the changing conditions of the postwar world, the immense increase in demand for oil and oil products, especially by Western Europe, the new petroleum reserves discovered by large-scale drillings in the Middle East, and the regrouping of some of the major international companies in that area, it became a pressing matter of policy for the major Middle East petroleum companies to arrive at an understanding among themselves. A petroleum agreement dealing mainly with problems of price-fixing on a worldwide market* was thus signed between major U.S., English, and French Middle Eastern petroleum companies in 1948, to the alarm of all those in the United States who wished to extend the meaning of the Sherman Antitrust Act to all domestic as well as international petroleum enterprises. The extension of the national jurisdiction over the activities of U.S. companies operating abroad had already been established in the case of U.S. vs. Socony-Vacuum (310 U.S., pp. 221-223), in which the court had declared that

> Any combination which tampers with price
> structures is engaged in an unlawful activity.
> Even though the members of the price-fixing
> group were in no position to control the mar-
> ket, to the extent that they raised, lowered,
> or stabilized prices they would be directly
> interfering with the free play of market
> forces. The Act places all such schemes be-
> yond the pale and protects that vital part of
> our economy against any degree of interfer-
> ence. . . .
> Under the Sherman Act a combination
> formed for the purpose and with the effect
> of raising, depressing, fixing, pegging, or
> stabilizing the price of a commodity in
> interstate or foreign commerce is illegal
> per se.[79]

In a later passage the court stated that purpose and effect were not essential to prove an unlawful restraint of trade. Because of jurisdictional issues the case could be treated

------

*The agreement, however, was mainly concerned with the European market.

as one where exertion of power to fix prices
. . . was an ingredient of the offense.  But
that does not mean that both a purpose and
power to fix prices are necessary for the
establishment of conspiracy under Section 1
of the Sherman Act . . . it is well established that a person "may be guilty of conspiracy, although incapable of committing the
objective offense."[80]

The court's opinion was also supported by the Webb-
Pomerene Act (40 stat. 516) to the effect that "American companies have been permitted to operate freely abroad in accordance with the economic and legal conditions there con-
fronting them, so long as the anti-trust laws of this country
have not been violated."[81]

In spite of the court's opinion and the evidence of the
Webb-Pomerene Act, the prevailing attitude adopted by the majority in the Senate was, as already noted above, that the
U.S. government could not enforce its laws beyond its own
territory.


## U.S. FOREIGN PETROLEUM POLICY DEFINED

The Senate Special Committee Investigating Petroleum
Resources had submitted its recommendations on January 31,
1947, but no official U.S. oil policy had resulted.  The subsequent agreement between the major Middle East petroleum
companies, in 1948, focused attention once again on the absence of a national petroleum policy.  The subject was in-
tensely discussed by the several branches of the government,
by the public, and by the petroleum industry.  The Department
of the Interior, which, through its Office of Oil and Gas,
had always been responsible for all government policies and
actions in respect to both domestic and foreign petroleum,
was especially concerned with the lack of such guiding principles.  In the absence of such guidelines, every decision
reached by the government was an isolated link in a series of
short-range and unrelated schemes devised only to offer temporary solutions.  Therefore, on July 3, 1948, Secretary of
the Interior A. J. Krug requested the National Petroleum
Council to draft a proposal for a petroleum policy for the
United States.

On January 13, 1949, the council submitted to Krug a
draft proposal for his approval.  This proposal may be regarded as the definitive national petroleum policy of the

United States. It defined the national interest in petroleum resources, both domestic and foreign; the relations
between government and industry; the long-range effects on
free market operations; recommended practices for conservation methods and goals; and described the relations of all
these elements to national security. The draft proposal endorsed the principle of a "vigorous oil development under
competitive conditions at home and abroad."[82] "The public
interest," the report of the council said, "can best be
served by a vigorous, competitive oil industry operating
under the incentive of private-enterprise."[83] As for the
role of the government in the petroleum picture, the council
established that "no government action specifically affecting the oil industry should be taken without proper regard
for the long-term effect and without consultation with the
industry," because, it added, "the oil economy is acutely
sensitive to governmental interferences with the free market."[84] The petroleum policy of the United States, the
council advised, should be based on the domestic production
of petroleum. Both "the national security and welfare required a healthy domestic industry" capable of providing "a
maximum supply of domestic oil to meet the needs of the
nation . . . under sound conservation practices." In saying that, the National Petroleum Council's report on national petroleum policy did not differ from the findings of
the Senate Special Committee Investigating Petroleum Resources. However, regarding the principles for U.S. foreign
petroleum policy, the council's recommendations were more
specific than those that were endorsed in the final report
of the Senate Special Committee. The council's report advocated that

> The participation of United States nationals
> in the development of world oil resources is
> in the interest of . . . our national secu-
> rity. . . .
>     Oil from abroad should be available to
> the United States to the extent that it may
> be needed to supplement our domestic sup-
> plies. The availability of oil outside of
> the United States, in places situated to
> supply our offshore requirements in time of
> emergency, is of importance to our national
> security. . . .
>     American interests today participate
> widely in international oil development.
> Conditions should be fostered that will

further this participation but not to the
extent that this involves preferential
treatment of operations abroad at the expense of the domestic industry.[85]

Along with this draft, the National Petroleum Council's Committee on National Petroleum Emergency also submitted, on January 13, 1949, its recommendations for a national petroleum policy for times of national emergency.
The report suggested two fundamental principles:

1. Oil facilities: most of the facilities of the petroleum industry including oil
wells, pipe lines, barges, tankers, refineries and distribution equipment, are usable
only in the petroleum industry and can therefore be co-ordinated on a vertical basis
without serious conflicts with other industries.
2. Government control: best results
can be attained if these controls are held to
a minimum and such as are necessary, worked
out in collaboration with the executive and
technicians of the industry.[86]

In view of the "many values of continuity," the committee recommended that in case of an emergency, "there should
be an organization of the petroleum industry similar to the
Petroleum Industry War Council and that there should be a
government agency similar to the Petroleum Industry War
Council and that there should be an Administrator."[87] The
committee further stressed the value of interagency committees in carrying forward a national emergency effort[88] and
suggested the inclusion in such committees of representatives from the Department of State, the Armed Services, the
Department of Justice, and the War Production Board.
Both these reports became the cornerstone of policy.
While the report of the National Petroleum Council underlined the general principles that were in turn endorsed by
the Department of the Interior for a national petroleum
policy, the report of the Committee on National Petroleum
Emergency provided the guideline for the organization and
conduct of oil crises. In these two reports are to be
found in a nutshell the principles that were later worked
out in practice during the following two decades of Middle
East oil crises. While the report of the National Petroleum
Council can be viewed as defining the limits of the voluntary

agreements under which the future Foreign Petroleum Supply
Committees were to operate, the report of the Committee on
National Petroleum Emergency can be regarded as applying
directly to the structure and responsibilities of the ad hoc
Middle East Emergency Committee of the 1956-57 oil crisis.*

## VOLUNTARY PETROLEUM AGREEMENTS

Once a national foreign petroleum policy had been de-
finitively formulated, U.S. decisions could be guided by its
principles.  However, the mechanism by which this policy was
to operate was not provided until, under the stress of the
emergency in Korea, Congress enacted on September 28, 1950,
only three months after the outbreak of hostilities in the
Far East, the Defense Production Act for the exploration and
development of mineral resources.  In this legislation,
Congress recognized that such work should be done by private
enterprise, but it conceded that where private capital could
not assume the long-term burden, it would be necessary for
the government to offer financial assistance, particularly
to industries dealing with mineral reserves that were in
short supply and critically needed for national defense and
civilian requirements.  The act provided "much of the basic
authority to bring about needed expansion of productive capacity, to provide controls over the use of scarce materials and to initiate other measures essential to enhance
the military strength."

Most important for the framework within which the international petroleum industry was to operate in time of
emergencies was Title VII of the act, which provided several
supplementary authorities.  In particular, of very great
importance was Section 708 of the act, which authorized the
president "to consult with representatives of industry,
business, financing, agriculture, labor, and other interests, with a view to the encouraging of the making by such
persons with the approval by the President of voluntary
agreements and programs to further the objectives of this
act."[89]

The petroleum industry, however, refused to join such
agreements for fear that while complying in doing so with

--------------------

*For further information on the voluntary agreements,
the Foreign Petroleum Supply Committees, and the Middle
East Emergency Committee, see next section and also appropriate sections in Part II.

the wishes of the Department of Commerce and the Department
of the Interior, it would be charged by the Department of
Justice for violations of the Sherman Antitrust Act.

It was not until the attorney general was persuaded
that the existing national emergency required the subordination of the Sherman Act to the requirements of national
security, that he agreed to accept the authorization of Section 708 of the Defense Production Act, which also granted
an immunity from prosecution for antitrust practices to all
parties joining such voluntary agreements. Accordingly,
one year to the day after the outbreak of hostilities in
Korea, on June 25, 1951, a Voluntary Agreement Relating to
Foreign Petroleum Supply was entered into by 19 U.S. international oil companies, the director of the Office of Defense Mobilization, and the attorney general, as a result of
which the first Foreign Petroleum Supply Committee (FPSC)
was established. There were five voluntary agreements relating to foreign oil supply between 1951 and 1967 (Table 3).

TABLE 3

Voluntary Agreements Relating to Foreign Oil Supply

| Voluntary Agreement | Duration of FPSCs* |
|---|---|
| a. temporary emergency | |
| agreement June 25, 1951 | June 26, 1951-July 8, 1952 |
| reactivated July 9, 1952 | July 9, 1952-Apr. 30, 1953 |
| b. basic agreement | |
| May 1, 1953 | June 24, 1953-Mar. 24, 1954 |
| c. amended Apr. 15, 1954 | Apr. 16, 1954-Mar. 25, 1956 |
| d. amended May 8, 1956 | Aug. 7, 1956-June 27, 1960 |
| e. amended Oct. 24, 1961 | Oct. 24, 1961-through the 1967 emergency; FPSC was enlarged at that time by (a) a "readi- ness" committee and (b) Se- curity Subcommittee |

*Having been created by the urgency of the current Ira-
nian oil crisis, the FPSC's first task was "to devise some
plan to help neutralize any major Iranian loss that might
occur." (The Iranian oil crisis is explored in detail in
Chapter 5.)

Source: U.S. Department of the Interior, Office of
Oil and Gas, The Middle East Petroleum Emergency of 1967
(Washington, D.C.: Government Printing Office, 1969),
Vol. II, appendixes, pp. A2-A6.

Soon it became apparent that the voluntary agreement
was the only available instrument by which the U.S. government could seek the cooperation of the U.S. Middle East oil
companies in matters of foreign policy and strategy.  And
it also became apparent that the good will of the petroleum
industrialists could be relied on only as long as they felt
reasonably secure from any more governmental encroachments
on their freedom of operation.*

However, the form as well as the application of the
voluntary agreement evolved in the course of the two decades
of oil crises.  As the political situation in the Middle
East became more and more precarious for the purposes of
U.S. economic enterprises, subsequent voluntary agreements
tended to tighten the cooperation between government agencies and Middle East petroleum companies, which now became
dependent more than ever before on U.S. foreign policy in
this part of the world.  The increased dependence of the
international petroleum industry on government policies in
the Arab countries contributed to a greater measure of cohesion and smoothness of operation between government and
industry within the framework of subsequent Foreign Petroleum Supply Committees, which, during the following two
decades, were chiefly preoccupied, in time of emergencies,
with the pressing problem of maintaining a continuous flow
of oil to Western Europe.

## SUMMARY AND CONCLUSIONS

U.S. oil activities in the war and immediate postwar
years may impress us today as being a muddled mixture of
confused thinking and undirected efforts.  Several simultaneous policies were undertaken that seemed to contradict
one another both in means and in goals.  While the Department of State was endeavoring to settle foreign oil distribution "equitably" by means of diplomatic agreements, the
Department of the Interior was attempting to establish U.S.
dominance in Arab oil fields.  At the same time, another
segment of U.S. officialdom and some members of the press
advocated the proposal of a UN commission in charge of all

---

*The new status quo between government and industry
was badly shaken by the Federal Trade Commission's severe
condemnation of the U.S. oil companies operating abroad in
its report entitled The International Petroleum Cartel
(1952).

raw materials, thus handing over all final oil decisions to an international organization.

In effect, however, all this hyperactivity in the various government departments was symptomatic of the time. The enormous resources of human energies that were released by World War II were expressed as much in immediate battle action as in grand designs. But these designs derived their impetus from traditional grooves of political thinking. They were, in fact, continuing trends that gained increased momentum due to the overall war effort. Thus, we can identify each government activity regarding Middle East oil with past experiences on the national and the international scenes.

The Department of the Interior's persistent efforts to impose U.S. government managerial dominance on all Saudi Arabian and Kuwaitian oil employed the same type of reasoning as had underlined the government-industry conflict in regard to domestic oil in the early New Deal era. Efforts to establish public ownership of foreign oil resources were not made to maintain the principle of local Arab sovereignty--as liberals fearing intentions of U.S. dollar imperialism had charged--but, instead, in order to dislocate the private sector from an area of decision-making that, in the opinion of the Roosevelt Administration, belonged in the hands of the elected representatives of the whole nation. Thus, the rivalry between public and private authorities in relation to Middle East oil was a carryover from an earlier Roosevelt administration. But in its dedication to the national welfare, the Department of the Interior failed to distinguish between the political properties of domestic and foreign petroleum resources.

In examining the efforts of the Department of State to solve the problem of foreign oil distribution, we again identify them with traditional diplomatic policies. Great Britain, the dominant power in the Middle East until that time, had several times arrived at international agreements with other European nations in regard to Middle East oil. Primarily, those agreements had been signed with the French government, but other agreements had also been ratified between Great Britain and Holland and, before World War I, the Imperial German government. Thus, when the U.S. government decided to step into Middle East oil enterprises and the British government acquiesced in this decision, it was only logical, according to established diplomatic practice, that the two governments should negotiate an international agreement.

The public and private segments in the United States that supported the proposal of a United Nations trusteeship of all raw materials in the world were not breaking any new ground either. U.S. sense of guilt for having failed to join the League of Nations and thereby, presumably, for being responsible for the Japanese invasion of China and the Italian occupation of Ethiopia was substantially augmented as a result of World War II. The desire to make amends for past mistakes gave birth to an almost fundamentalist faith in the powers of international organizations to rectify all the evils of our times. Thus, a movement was born that claimed that all matters concerning raw materials of the world ought to be handled by an appropriate United Nations agency. Cold War winds, however, quickly extinguished its enthusiasm, and the proposition that Middle East oil be handed over to the United Nations rapidly became the political property of communist parties and front organizations.

In searching for a U.S. foreign oil policy, both executive departments and the legislature were dodging the central issue: the lack of a defined policy-making authority to deal with foreign oil problems. The Department of the Interior was obviously not the proper agency in which such decisions could take place. While it possessed a clear notion of the domestic interest, it had not enough understanding of U.S. relations with other nations. Its endeavors to step into Middle East oil enterprises violated every accepted international practice and were made in total ignorance of U.S. diplomatic relations and foreign interests and responsibilities in this region.

The State Department itself seemed to be allowed very little say in matters of Middle East oil, as all questions of energy supply were deemed to be in the province of domestic affairs. At the same time, the scope of the oil activities of both departments was severely restricted by existing antitrust legislation. Because of all these restrictive limitations on government powers, and in the absence of a defined decision-making authority in relation to foreign oil, it is not surprising at all that U.S. foreign oil policy was finally articulated by the private sector, namely, the private oil industrialists as represented in the National Petroleum Council.

In conclusion, then, it is quite clear that unless the United States defines a public authority equipped to deal with foreign oil problems, it will continue to be the victim of circumstances and its vision of the problems involved in foreign oil in general and Middle East oil in particular will continue to be obscured by a host of irrelevant issues.

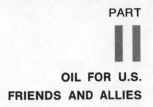

# 4

## THE POSTWAR YEARS

The cornerstone of much of U.S. foreign policy is that a secure, democratic, and prosperous Western Europe is the bastion guaranteeing the peace and security of the Western Hemisphere. To maintain Europe in that position of strength, the United States spent billions of dollars, and this process culminated in the enunciation of the Marshall Plan and the signing of the NATO Alliance. The spectacular success with which Europe met its energy crisis in the postwar world was one of the early signals of the continent's imminent recapture of its traditional place in the world economy.

This chapter analyzes the role of petroleum in the postwar recovery of Western Europe. It compares the pessimistic views of the United States about European fuel shortages, discussed in the first section of this chapter, with the resolute European approach to its energy problems and their solution, described in the second section.

## U.S. FOREIGN AID AND WESTERN EUROPE'S PETROLEUM REQUIREMENTS

In Western Europe, the new postwar economy of the recovering continent was becoming increasingly dependent on the availability of petroleum. Unlike other forms of assistance received from the United States, it was evident that due to U.S. petroleum policy, the European nations could not depend on U.S. oil. On July 16, 1947, Congress passed a bill (H.R. 4042) to control the export to foreign countries of gasoline and other petroleum products from the United States. The Senate report that accompanied the bill provided that "no diesel oil, bunker oil, or lubricating oil" would be moved or transported from the United States to a foreign nation "unless the Secretary of Commerce shall certify to the President that such movement or transportation will not impair the national defense, endanger the national security, nor impair the civilian use of these products by the people of the United States."[1] The report further concluded that the current shortage was due in part to the policy "of permitting large-scale petroleum exports. Our view is that the needs of American consumers must be placed before the requirements of foreign nations."[2]

Concerned about the future of European recovery, both for strategic as well as economic reasons, the Truman Administration requested and received information on the question of petroleum supplies to Western Europe from three sources:

1. The Krug report on National Resources and Foreign Aid came to the conclusion that "the primary need for rebuilding Europe's petroleum economy was transportation capacity to carry Middle East crude oil to European refineries, the capacity of which was being increased by reconstruction and the building of new plants."[3]

2. The Harriman report on European Recovery and American Aid, more pessimistic than the Krug report, concluded that petroleum supply "would be a limiting factor because of the limitations on world production, refinery capacity and means of transportation."[4]

3. Finally, the Herter report on Petroleum Requirements and Availabilities, produced by the House Select Committee on Foreign Aid under the leadership of Representative Christian A. Herter, dealt in its Report No. 5 with petroleum supply and demand. This report was as pessimistic as the Harriman report inasmuch as it concluded that existing shortage of petroleum equipment, transportation, and refineries made it impossible to meet Europe's need for civilian petroleum supplies.[5]

All three reports agreed that both refinery capacity and tanker-building production in Europe should be increased in order to maintain the continuous flow of adequate amounts of petroleum to the European nations. All three reports agreed that petroleum supplies to Europe should be forthcoming from Middle East sources. This view was in accord with the general tenet held in Washington that domestic reserves were limited and therefore not exportable.

In addition to these reports, the Department of State included a report on petroleum in its Outline of European Recovery Program, which it prepared for the Congressional hearings on foreign aid.[6]

The results of these diverse efforts were the Foreign Assistance Act of 1948[7] and the related Foreign Aid

Appropriation Act of 1949. Under the heading of "Protection of the Domestic Economy" (Section 112) of the Foreign Assistance Act, it was stipulated that, in consideration of then current and anticipated world shortages of petroleum and its products, petroleum for Europe must, to the maximum extent possible, be procured from sources outside the United States.*

Under the Marshall Plan, the participating European countries set up the Organization for European Economic Cooperation (OEEC), which, in turn, appointed a Fuel and Power Committee consisting of two sections: a Petroleum Subcommittee and a Coal and Electricity Subcommittee. The so-called Paris report, dated September 22, 1947, of the Petroleum Subcommittee analyzed the causes of the dollar shortages in the OEEC member countries and made recommendations for the Economic Cooperation Administration (ECA) Act of 1948 to the effect that, whenever possible, petroleum for Europe should be purchased elsewhere than in the United States.[8]

Expected shortages of available petroleum reserves in the United States, then, prohibited the U.S. government from providing petroleum supplies for the postwar recovery of Europe, and dollar shortages within the OEEC area dictated a similar policy of diverting European petroleum procurement away from the United States. Europe's fate was thus sealed for a substantial dependence on Middle East petroleum sources for its energy supply. Europe's recovery was henceforth inconceivable without a constant flow of those readily available crude petroleum supplies from the Arab countries and Iran.


WESTERN EUROPE'S PATH TO RECOVERY

As seen above, three well-thought-out reports advised the president of the United States that Europe's economic recovery was a problematical matter at best. The Department

---

*Their exclusion from the lucrative European market subsequently caused a great deal of furor among the domestic producers of petroleum in the United States, who pointed to the fact that while American petroleum production had increased only 8 percent from 1947 to 1950, Middle East production had increased by 109 percent, charging that "seldom has a Government policy so clearly resulted in abolishing a foreign market for a domestic industry." (See Elmer Patman, The Third World Petroleum Congresses [Washington, D.C.: Government Printing Office, 1952], p. 7.)

of State's commodity report regarding petroleum positively
predicted that "less essential uses in Europe would continue
to be restricted and that the projected large-scale coal-to-
oil conversion program would not be carried out."[9]  In the
U.S. view, then, difficulties of oil supplies to Europe were
tremendous because of the continent's shortage in refinery,
tanker, steel, and other programs.

Given this view, the OEEC Petroleum Committee could not
even hope to attempt to start serious planning on such premises.  It simply assumed, then, that "the estimated European
requirements would in fact be met; no attempt therefore was
made to relate these requirements to a possible world availability over the period."[10]  The extraordinary results of
this economic leap into the future in regard to petroleum
soon became apparent.

Defying all official U.S. economic reports and expecta-
tions, Europe's pace of economic recovery was perhaps the
most extraordinary event of the postwar years.  In the face
of U.S. predictions to the contrary, Europe started to re-
habilitate its domestic industry and to recapture its former
place in the world market.  Nowhere in the postwar world was
the essentiality of oil to contemporary civilization so remarkably demonstrated as in Western Europe.  Oil played a
major role in European recovery.  Without oil the economic
life of the OEEC countries, their technical and industrial
progress, and the promotion of the individual standard of
living would have been seriously impeded.  However, the prewar role of Europe as an importer of finished products under-
went a major change.  Europe's need to minimize the expenditure of foreign exchange led to an emphasis on the importation of crude oil rather than the more expensive finished
products.  (See Table 4.)  That, in turn, led to the development of a major refining industry in the OEEC area, which
was materially aided by the Marshall Plan.

The rapid growth of the European refinery industry reflected the paramount need to save foreign exchange.  Because of this need, the purely technical oil economics was
subordinated to balance-of-payments considerations.  The
European program of refinery expansion inevitably reflected
Europe's current difficulties with balance of payments.[11]
Soon Europe achieved one of the main purposes of the postwar
expansion of refineries, namely, the saving of foreign exchange, particularly dollars, as illustrated in "the reduction of its dollar liabilities both for imports of petroleum
and for equipment."[12]  At first, the European refinery program provided for what were essentially the domestic demands

TABLE 4

Crude Oil and Finished Products, OEEC Countries,
Prewar, 1948/49, and 1952/53
(millions of tons)

|  | Prewar (Average) | 1948/49 (Actual) | 1952/53 (Est. Dec. 1950) |
|---|---|---|---|
| Refinery throughput | 12.0 | 23.0 | 65.1 |
| Crude oil supply Imports: | | | |
| Dollar | 6.4 | 9.2 | 12.9 |
| Nondollar | 4.6 | 12.4 | 48.2 |
| Indigenous production (including shale oil) | 0.7 | 2.0 | 4.0 |
| Refinery output | 11.2 | 20.9 | 59.2 |
| Net imports | | | |
| Dollar | 7.3 | 11.6 | 5.7 |
| Nondollar | 13.2 | 11.9 | 5.7 |

Source: Organization for European Economic Cooperation, Oil Committee, First Report on Coordination of Oil Refinery Expansion in the O.E.E.C. Countries, Report by the Oil Committee on the Long-Term Plans of the Participating Countries (Paris, October 1949), pp. 33, 35.

of the OEEC member states.* But in a very short time, the expansion of European refining capacity reached the point where substantial exports were being made to points beyond the participating countries.[13] The principal exporting countries were those that also had the largest European petroleum companies operating in the Middle East--namely, the Netherlands, the United Kingdom, and France. But other countries, especially Italy and Belgium, already began to

---

*That was the general rule to which France provided the exception, as the French government's consistent aim was to supply a large part of their overseas territories from French refineries.

develop a refinery capacity that, in turn, cultivated substantial exports of petroleum products to their credit. Such refineries as Naples and Shell Pernis were among the earliest in Europe to be planned for export. The existence of a major European petroleum refining industry made it possible to develop other industries in Europe. (See Table 4.) As a result of the postwar refinery construction, Europe established a growing number of petrochemical plants that often worked in association with the refineries.[14] Thus, the acceleration in the buildup of a refining capacity in Europe was motivated by a combination of the following: an alleviation of the balance-of-payments difficulties; a general tendency to relocate refineries from points of production to points of consumption; a rapid upward trend in consumption in all the OEEC countries; and a convenient outlet for the Middle East companies to market their oil supplies.[15]

Technically, the rapid transition of European industry from coal to oil, which took place during the postwar years, was possible only as long as a reliable fleet of tankers operated between the Persian Gulf states and the European harbors, where the oil was then pumped and shipped to a growing network of European refining facilities to be resold in the form of various petroleum and petrochemical products of multiple applications.

It was at this point in the phenomenally rapid recovery of Western Europe that the continent was suddenly hit by the Iranian oil crisis, the shutting off of the Abadan refineries, the boycott on all Iranian oil by the major Western companies that ensued as a result of the expropriation of the Anglo-Iranian Oil Company, and, finally, the general stoppage of normal petroleum flow from Iran. The entire European economy was thus threatened.

The problems raised for Europe and the United States as a result of the crisis, the solutions they received, and the implications they bore for future U.S. foreign policy are discussed in the next chapter.


SUMMARY AND CONCLUSIONS

During World War II, a conviction grew in the United States that U.S. failure to join the League of Nations after World War I and its refusal to guarantee the French frontier[16] had led many countries to believe that this government "had no interest, and would not seriously concern herself, in what was happening in Europe, in Africa, or Asia;"[17] and that because of this belief Germany had adopted an aggressive foreign policy, which had led, ultimately, to war. This theory of U.S. responsibility placed upon U.S. shoulders a burden of guilt resulting in a consuming desire to rectify past omissions. Thus, Secretary of State Byrnes could inform his European colleagues that "America is determined this time not to retreat into a policy of isolation. We are determined this time to cooperate in maintaining the peace."[18]

U.S. desire to assist Europe in its path to recovery was justified by reasons of economic prerequisites for a profitable international trade and domestic prosperity. It was widely maintained in the United States that "only by economic recovery of other countries" could the U.S. people "hope for the full employment of our labor and our capital in this interdependent world."[19] These arguments were soon reinforced by additional considerations of national security generated by the Cold War. Thus, the "traditional policy of 'going it alone' was to be replaced by a policy of 'going it with others'."[20] A U.S. policy of worldwide alignments replaced the former policy of isolationism.

Not in every case, however, could U.S. foreign policy respond to European commodity needs. Domestic considerations were bound to override foreign policy objectives. The State Department's aid programs to European recovery had to be reexamined in light of domestic production requirements. Thus, domestic policies placed important material curbs on foreign policy decisions. Europe's need for petroleum and petroleum products could not be met by the United States either as outright sales or as long-range credits, since the Oil and Gas Division of the Department of the Interior considered the preservation of all domestic reserves essential for the national welfare. Even after the oil scare of the mid-1940s had subsided, there remained the problem of replenishing the oil reserves of the nation in accordance with the country's conservation policy.* Such reserves had been drained by the war effort and had now to be filled up. Because of the need to build up these oil reserves, no assistance was forthcoming from the U.S. government to the European nations in acquiring the single most basic commodity necessary for the recovery of their industry, the fueling of their homes and plants, and the general rehabilitation of the whole continent.

Instead of relying on U.S. planners, then, the European nations had to devise themselves the means for their own

_____

*For U.S. conservation policy, see appropriate section in Chapter 7.

survival. It was evident from the beginning of the postwar era that only by innovative institutional structures could the European economy become a healthy one. These new structures required a central integrative machinery that would gather information and dispense materials. By withdrawing its material aid to Europe in several essential areas and confining it to others, secondary commodities as well as financial cooperation, the United States, _ipso facto_, excluded the possibility of creating the economic and political machinery that could provide the means toward an Atlantic partnership. Even if it is true that "it is the monetary factor that creates the strongest element of interdependence,"[21] this element is only one among many other elements contributing to interdependence and its intergovernmental machinery is generally such that it is one of the least important factors contributing to regional integration.

The inability (or unwillingness, as the case may be) of the United States to enter an economic organization encompassing all of NATO's members caused a sharp division of purposes in the political evolution of Western Europe. While security considerations drove Europe to look to the United States to take a leading role, the separate economic evolution of the OEEC thrived on exactly the opposite theme, namely, the strongly particularist interests of the area. Thus, very early after World War II, the political future of Europe was bifurcated: On the one hand, NATO was blamed by the European purists for "Atlanticizing" the continent; on the other hand, those seeking closer ties with the United States blamed the purists for contributing toward a schism in the Atlantic partnership. The three major objectives of Western integration—namely, cohesion, security, and prosperity—were thus divided between two unequal international organizations. This initial dichotomy soon resulted in a multiplicity of schemes for European unity. On one thing, however, all these diverse advocates agreed, and that was that Europe must regain its former position in world affairs. Thus, U.S. absence in the economic organizations of the continent, which in the early postwar years had seemed to forebode adverse effects upon the course of European recovery, produced exactly the opposite result: Because an economic partnership across the Atlantic Ocean was not forthcoming, Europe was driven very early in its progress toward recovery to rely on its own resources. While in matters of security Europe showed the strongest dependence on the United States, in economic matters it was rapidly becoming the master of its own destiny. The future of Western relations now depended on a balance between European security and European economy.

CHAPTER

# 5

## THE IRANIAN
## OIL CRISIS

U.S. involvement in Iranian affairs began, in a very limited way, during World War II. It stemmed from the Soviet-Iranian dispute, a dispute ostensibly over oil concessions but, in reality, one that embroiled the British and U.S. governments as well. After the war, Washington remained apprehensive in regard to Soviet-Iranian relations. The creation of a "popular democratic republic" in the northern province of Azerbaijan, the continued presence there of Soviet troops, and the increasing influence of the communist Tudeh Party in Iran became central issues of interest to the Department of State and the U.S. press. The Truman Doctrine, which came about as a result of conditions in Greece and Turkey, also enunciated a general policy of containment of communism. Thereby, for the first time, an official document publicly admitted U.S. political interest in the Middle East as a whole and offered to assist these peoples in their resistance to "outside pressures."

In 1950 an oil dispute between the government of Iran and the Anglo-Iranian Oil Company (AIOC) was fanned by deep antagonism in Iran toward all foreigners but chiefly toward British and U.S. advisers and business concerns. Although the U.S. government continued its hands-off policy toward Iran even when U.S. concerns and missions were requested to leave the country (primarily because Iran was considered at the time to be within the British sphere of interest), it was not until Great Britain had severed relations with Iran as a result of the expropriation of the AIOC that the United States actually intervened in the dispute, this time in the role of a mediator.

This chapter describes, first, the political background in Iran, which contributed toward the oil crisis of 1951-53. The chapter's second section describes the role the United

States played in alleviating the oil shortages created by the interruption of oil exports from Iran. The third section describes the influence of the oil crisis on the European oil industry, and the fourth deals with the organization of the Iranian oil consortium.

## THE EXPROPRIATION OF THE AIOC'S PROPERTIES IN IRAN

In the early 1950s, Iran became again a priority problem area of the Western powers. Situated on the southern flank of the Soviet Union, located on a highly strategic route toward the Indian Ocean, and containing enormous resources of oil as well as the world's largest refinery at Abadan, Iran soon began to occupy the center stage of international relations. With a population of 50 million people, 90 percent of whom were illiterate and 80 percent farmers (90 percent of these farmers were working for absentee landlords), the country seemed to offer a fair target for communist propaganda. The Tudeh Party* promised land reform and launched virulent attacks against all British and U.S. oil interests in the country, which, it said, included "military, security, missionaries, businessmen and technical advisers."[1]

At the same time, 1950 was, in Iran, a year of economic crisis and depression. Inflationary conditions impelled the government to postpone its seven-year plan based on a $100 million budget. The depressed conditions spurred a large-scale flight of capital from the country and, consequently, forced the authorities to place severe restrictions on imports. Up until that year, about one-half of all Iranian imports had been purchased in the United States. In 1950 there was a marked shift of orders toward the sterling area. Whether because of this shift or for political reasons, U.S. promises of loans and grants to Iran were not honored. The only money successfully contracted by the government of Iran that year, although not actually received until 1953, was a $25 million loan extended by the Export-Import Bank.[2] A loan application made by Iran to the International Bank for Reconstruction and Development

---

*The Tudeh Party (the communist party of Iran) attempted to assassinate the Shah on February 4, 1949; the following day, the party was outlawed and its leaders were arrested. On December 15, 1950, many of the top leaders managed to escape and began to reorganize the underground Tudeh Party and to press for its legalization.

was rejected, with the result that the embittered Iranian government threatened to cancel its membership in the bank.[3] The rejection of Iran's loan applications gave the National Front in the Majlis* an opportunity to press for radical revisions in the oil concession agreement held by the British company.

Iran was the fourth country in the world in oil production and the third in estimated reserves. The largest and only substantial concession-holder in the country was AIOC. The company was the chief supplier of fuel oil ("bunker") to the British Royal Navy. It owned the world's largest refinery, at Abadan, and a pipeline network that brought the oil from the oil fields in Iran to ports on the Eastern Mediterranean, where it was pumped into fleets of tankers. The company had aggregate shares on the London stock market in the amount of about $720 million, and its actual worth was far greater.†

The original AIOC concession was first granted on May 28, 1901, to William Knox D'Arcy, a British subject, upon the liquidation of Baron Julius de Reuter's concession (1872-1901) for a 70-year multiple operations monopoly. The D'Arcy natural gas, petroleum, asphalt, and ozocerite concession was granted for a period of 60 years in exchange for £20,000 in stock and 16 percent of the annual net profit.

The British government became actively interested in Iranian oil when on October 21, 1904, Admiral John Arbuthnot Fisher was appointed first sea lord of the Admiralty. Fisher immediately appointed an oil committee, which began to implement the belief he had held since 1882 that the Royal Navy ought to convert from coal to fuel oil. In 1908, at the suggestion of the Admiralty, the Burmah Oil Company, which up until then was the chief oil supplier for the Royal Navy, formed a syndicate with D'Arcy in order to develop the Persian concession. The new merger became known as the Anglo-Persian Oil Company. Lord Strathcona of the Burmah Oil Company was made chairman, and D'Arcy became a member of the board of directors. In 1910 Fisher resigned and was succeeded in 1911 by Sir Winston Churchill. Churchill, who was deeply impressed by the Agadir incident of that year, persuaded Fisher to come out of retirement in order to assume the chairmanship of the commission on oil supplies, and it was in this commission that the decision was made

---

*The Iranian parliament is made up of two houses: the Senate and the Majlis, which is the house of representatives.

†The Iranian government, however, claimed the worth of the company to be $585 million or less.

according to which in June 1914 the British government became a joint owner of the Anglo-Persian Oil Company, holding a controlling interest of 53 percent for a consideration
of £2.2 million.

On April 29, 1933, after long negotiations with the
Iranian government, a new concession agreement was signed
between the government of Iran and the Anglo-Persian Oil
Company according to which royalty payments were placed on
the basis of an involved tonnage computation, thus avoiding
the uncertain fluctuations of posted prices on the international market. The new concession was to run until 1993.

The new agreement that was negotiated between the Iranian government and the company in 1949 provided for a minimum of $61 million in annual revenue and nearly double the
ratio in royalties. However, this agreement coincided also
with a mounting feeling in Iran against all foreigners and,
consequently, against all concessionary agreements with foreign concerns. Moreover, the international environment, the
recent war in Greece, the current war in Korea, and the increasing tensions generated by the Cold War produced a fear
in Iran that it was being used as a pawn in the historical
British-Russian rivalry in the region. The National Front
claimed that this agreement was highly unfavorable to Iran
as compared with concessionary terms in other producing
countries. In Venezuela, for example, the U.S. concessionary had recently agreed to pay the government a royalty of
50-50 percent of the annual net profit. Venezuela had thus
set a precedent and an example to other producing nations.
Immediately Saudi Arabia and Aramco had signed a similar
agreement, whereupon reformers in the Iranian Majlis began
pressing for a similar revision in the Anglo-Iranian agreement, while Dr. Mohammed Mossadegh, leader of the National
Front, and his entourage began clamoring for the nationalization of the oil industry. In the face of the demand made
by both nationalists and communists to expropriate the company's properties, the 1949 agreement between the Iranian
government and AIOC was never ratified by parliament. On
December 26, 1950, the Oil Committee forced the Majlis to
back out of the agreement on the grounds that "the proposed
increase in royalties and of taxes was not enough to safeguard Iranian interests."[4]

At the same time, the Iranian government cancelled the
contracts of 11 U.S. managerial concerns grouped together
under the name of overseas consultants and serving as advisers in connection with the intended seven-year plan for
economic development. The Iranian government claimed that
these U.S. concerns had made important appointments "on the

basis of political and personal interests instead of competence and experience" and that the "utilization of money available was determined by personal and political interests." It also accused the overseas consultants of graft and unscrupulous manipulations of its accounting books.[5]

Thus, a fertile ground was prepared for conditions of political unrest. Nationalist groups, demanding the ouster of all foreigners from Iran and the expropriation of AIOC properties, found allies in the Tudeh Party, which was skillfully directing mounting public resentment toward its own ends. The emergence of the communists in Iran was viewed with alarm in both Great Britain and the United States, and it was noted that during the last six months of 1950, Iran, in addition to expelling overseas consultants, had also banned the broadcasts of the Voice of America and the British Broadcasting Corporation, while, at the same time, it had signed a number of trade agreements with the Soviet Union. Viscount Jowitt, lord chancellor of Great Britain, warned in the House of Lords against the self-deception that what was taking place in Iran was not in fact an evidence of Soviet aggression "under the guise of domestic insurrection."

> It is not impossible that those who preach
> world revolution will seek to advance the
> borders of their empires by starting what
> appear to be civil wars but are in fact
> aggression in disguise. They will hope
> that the free world will be confused and
> divided in its opinions as to the actions
> it should take. The Government is well
> aware of this danger and will not be easily
> misled by anything that may happen, be it
> in Central Europe, the Middle East or fur-
> ther afield.[6]

U.S. and British apprehensions were aggravated when, on March 7, 1951, Premier Ali Razmara, an anticommunist whose reform program was backed by the United States, was assassinated in a Teheran mosque by a member of Fadayan Islam* following his 7,000-word report "bitterly opposing the nationalization plan on the ground that it should take

---

*Fadayan Islam ("Crusaders for Islam") demanded an end to all foreign influences in Iran and the return to the strict principles of the Koran.

away four-fifths of Iran's revenues and throw thousands of persons out of work."[7] The Fadayan Islam also warned that all leaders voting in favor of Razmara's report would meet a similar fate. Razmara believed that investment capital for the future economic development of Iran could be obtained only from oil and that this capital could only come from abroad. In the United States, Razmara was regarded as the strongest supporter of long-range U.S. policy in the Middle East, and his violent death was felt as a blow to U.S. interests.

Upon Razmara's assassination, Shah Mohammed Reza Pahlevi immediately named 70-year-old Khalil Fahimi to be the acting premier, pending parliamentary approval, and assigned him the task of forming a caretaker government. In naming Fahimi, the shah was, in fact, affirming the same conviction that had led him to appoint his predecessor-- namely, the need for a "strong man" who had also a clear anti-Soviet record. Razmara had put down the pro-Soviet rebellion in Azerbaijan in 1946, and Fahimi had been governor of this northern province, which borders on the Soviet Union. It was expected that the new premier would implement Razmara's policies. But on the day following the assassination, a parliamentary oil committee voted unanimously (15 of the 18 members were present) in favor of a recommendation for a countrywide nationalization of all oil resources. The bill was approved by the Senate on March 20, 1951, a date that thus became the official anniversary of the nationalization of Iranian oil resources. At the same time, the Majlis rejected the shah's appointment of Fahimi. In his place the shah now appointed the former Iranian Ambassador to Washington, Hussein Ala. Like his two predecessors, Ala also had an anti-Soviet record, as he had handled the Iranian case in the United Nations in 1946 concerning the evacuation of all Soviet troops from the northern province of Azerbaijan. Consequently, Western diplomats in Teheran regarded Hussein Ala as a good choice to resist communism in Iran. The nomination of Hussein Ala was unanimously approved by the Senate but evoked a storm of protest by the pro-Soviet bloc in the Majlis. When he was nevertheless confirmed by a majority in the Majlis (69 to 27, with 10 abstentions), members of the National Front walked out of the chamber.

At this point, the AIOC came forward with the offer of a 50-50 royalties share with the Iranian government along the same lines as the agreements made in Venezuela and Saudi Arabia. But the British, it seemed, had waited too long: They offered too little too late, for neither

the Majlis nor the Iranian public were now prepared to accept anything short of complete expropriation of all AIOC's
assets in Iran and the nationalization of all oil resources
in the country.  The British government reacted with a formal note to the government of Iran protesting any such plan
on the grounds that the AIOC had a firm agreement with the
government of Iran valid until 1993, and any unilateral
termination of such an agreement was illegal per se.  Moreover, the British note mentioned four major obstacles for
nationalization--namely, that (1) Iran did not have enough
money to compensate the British company for expropriating
its assets; (2) Iran did not have enough money to run the
operations; (3) Iran did not have enough trained personnel
to work in the operations; and (4) Iran did not have the
worldwide sales organizations and tanker fleets needed for
the distribution of oil and oil products.

The British note hardly made a ripple.  Demonstrations
and riots in favor of nationalization persisted throughout
the country.  The government declared martial law.  Tanks
and troops were in control of Teheran.  The riots spread to
Abadan and all other oil areas.  All communication between
Teheran and these parts was cut off.  In the midst of the
unrest, the Soviet press made accusations that added more
fuel to the already explosive situation.  On March 18,
Pravda accused the United States on its front page of having engineered the assassination of Razmara in order to direct Iranian public opinion against the Soviet Union,[8] and
in the same month the Literary Gazette (Moscow) charged
U.S. Ambassador Henry F. Grady with the scheming of a blood
bath for Iran and turning that state into "another Greece."
It called Grady a "specialized expert executioner and a
stifler of freedoms of people."[9]  Soviet special interest
in Iran, Pravda asserted, stemmed solely from the legally
binding sixth article in the Soviet-Iranian Treaty of 1921,
which said that in the event of any third party attempting
to carry out an aggressive policy in Iran or to use Iran's
territory as a place d'armes for military attack against
the Soviet Union, Russia has the right in the event that
Iran is unable to protect itself against such danger to in-
troduce Soviet troops into Iran in order to take necessary
measures in the interests of self-defense.

Mounting pressures by Majlis groups and increasing
riots throughout the country led on March 27 to the resig-
nation of Ala.  On April 28 the shah, upon the Majlis's
recommendation, reluctantly entrusted Mohammed Mossadegh,
leader of the National Front, with the formation of a new
cabinet.  Premier Mossadegh was known as one of the most

outspoken advocates of oil nationalization, but he was also
a stanch opponent of "all foreign aid including assistance
by the United States military mission to the Iranian army."[10]
It was immediately noted in Washington that Mossadegh granted
the first diplomatic conference after his appointment to the
Soviet Ambassador (May 1), and only the following day, on May
2, did he hold a conference with U.S. Ambassador Grady.

Mossadegh's first important act was to put into force
the recently enacted nationalization law of all Iranian oil
resources. As soon as the AIOC's properties in Iran were
nationalized, the British government severed its diplomatic
relations with the government of Iran. British applications
to the International Court of Justice for a judicial decision were rejected when the court declared itself incompe-
tent to deal with this case, as it was a matter of domestic
jurisdiction of the state of Iran.

Company ties and diplomatic pressures succeeded in en-
listing a more immediate action on the side of Great Britain.
U.S. and Dutch companies, fearing that Iran's example might
be followed by other oil-producing nations seeking to increase their control over their own oil resources, united in
a general boycott of all Iranian crude oil and oil products.
No company tanker would transport these resources from Iran,
nor would any distributing agency buy them for world marketing purposes. Thus, despite all its natural wealth, Iran
was unable to make any profit out of it. At the same time,
consuming nations, chiefly in Western Europe, were left to
fare without their supply from Iran.

As soon as Great Britain severed relations with Iran,
the U.S. government entered actively into the Iranian oil
picture. If U.S. diplomatic efforts did not help to repeal
the nationalization law in Iran and return the AIOC to its
former position in Iranian oil operations, they certainly
succeeded in opening the door for U.S. oil companies to
share in the spoils of Iranian resources.


## THE UNITED STATES IN THE IRANIAN OIL CRISIS

The year 1950 brought two closely linked events: The
Korean War broke out in the Far East, and a major Iranian
petroleum crisis occurred in the Middle East. Before the
Korean outbreak, there had been a practical balance between
gasoline demand and aviation gasoline supply throughout the
world. As soon as the demand went sharply upward because
of military buying, aviation gasoline became a scarce commodity all over the world. Most of this supply of aviation

gasoline was estimated to come from the Abadan refineries
in Iran.  The Abadan refineries were also the greatest single supplier of residual fuel oil ("bunkers") for ship operation.  Fuel oil from Abadan fueled most of the merchant
vessels and navies of the Eastern Hemisphere.

In April 1951, a two-week strike in the Abadan refineries of Iran deprived the Western nations of approximately
7 million barrels of petroleum.  According to the Department of the Interior,

> Before the Anglo-Iranian controversy reached
> its crisis peak in June 1951, Iran supplied
> more petroleum than any other country in the
> Middle East.  It accounted for approximately
> 7 per cent of the Free World's petroleum sup-
> plies, including more than one fourth of all
> refined products supplied outside the Western
> Hemisphere to the nations of the Free World.
> Crude oil in Iran approximated 660,000 barrels
> per day, more than one third of all Middle
> Eastern production.[11]

The Abadan refineries produced 18,400 barrels of aviation gasoline per day, amounting to 40 percent of the total
aviation gasoline production in "friendly foreign nations."[12]
510,000 barrels per day of crude oil were processed at Abadan's refinery.  Approximately 460,000 barrels per day (b/d)
of refined products were exported to friendly nations, as
follows:

| | |
|---|---|
| Aviation gasoline | 18,400 b/d |
| Kerosene | 46,000   " |
| Residual fuel oil | 210,000   " |
| Gas/diesel oil | 95,500   " |
| Motor gasoline | 87,200   " |

The strike threatened to put large segments of the air-
line industry at a virtual standstill.  At the same time, a
worldwide oil panic was followed by large-scale hoarding of
petroleum and petroleum products, causing chaos abroad and
disrupting U.S. domestic industry.  European oil companies
operating abroad became deeply concerned about their prospects in the oil-producing countries of the Middle East
within the immediate foreseeable future.  All these speculations increased many times when the AIOC's properties in
Iran were nationalized, and a general boycott by the international oil industry on Iranian oil deprived the world of

its supplies.  Thus, U.S., British, and Dutch companies deprived Western Europe of its energy supplies for reasons of self-interest in protecting their own commercial investments.  At the same time, nations like Italy that attempted to import oil from Iran on their own failed to do so successfully for lack of adequate organization, primarily in the availability of means of transportation.

The loss of Iranian petroleum created a new demand on U.S. refineries and thus wrecked the delicate balance between supply and demand in the United States.  Moreover, the stoppage of petroleum flow from Iran to all parts of the world threatened to undermine the military operation in Korea and to jeopardize the entire mobilization program of the United States.  The war machinery, being heavily dependent on Iranian oil products from the Abadan refinery, could have suffered a serious setback.

When Iranian oil ceased to flow and aviation gasoline from the United States and from South and Central America had, therefore, to be sent all over the world in ships to keep the world's fleet of airplanes flying, a solution was offered in the mobilization of the oil companies operating abroad.  Under the provisions of Section 708 of the Defense Production Act of 1950, a voluntary agreement was reached between government agencies and the petroleum industry. After setting an equitable scale of oil rationing, the companies were required to increase their production in order to meet these obligations.  However, only U.S. oil companies operating abroad could be reached directly in this way by the U.S. government.  Foreign oil concerns whose cooperation was also sought could be reached only through the diplomatic channels of the State Department.

In order to coordinate these worldwide operations, a Foreign Petroleum Emergency Committee was created made up of representatives of the oil companies and government officials.  Whatever objections the Department of Justice might have raised against such a commercial cooperation between government representatives and major oil corporations* was

---

*During World War II, Attorney General Francis Biddle objected to the participation of representatives of the petroleum industry in the negotiations concerning the proposed Anglo-American Petroleum Agreement on the grounds that such cooperation between government agencies and the private sector violated antitrust legislation.  At that time the Justice Department's ruling was respected by the U.S. government inasmuch as the American delegation to these negotiations, unlike the British delegation, consisted exclusively of government officials and military personnel.

taken care of by the Petroleum Administration for Defense in a meeting with Graham Morris, the assistant attorney general, in which he was told of the effect on the entire world economy unless means were enlisted to supplement shortages created by the shutdown of the Abadan refinery due to the general boycott by the Western oil companies. Representatives of the Petroleum Administration for Defense persuaded the assistant attorney general that "practicality had to prevail over legal precision." This advance clearance from the Justice Department was necessary in order to assure the oil corporations that they would not be prosecuted later on charges of collusion. After a series of conferences between the Defense Production Agency, the Federal Trade Commission, and the Justice Department, an agreement was finally approved.

Thus, a voluntary agreement, as provided for by the Defense Production Act, was reached between the U.S. government and the U.S. petroleum companies operating abroad. Although other voluntary agreements had already existed under the provisions of this act, this was the first voluntary agreement that was worldwide in scope. As soon as the voluntary agreement had been approved, it was set into operation through a network of committees that worked in concert under the direction of the General Committee of the Foreign Petroleum Emergency Committee, which, in turn, consisted of five subcommittees. Stewart P. Coleman, previously director of the Program Division of the Petroleum Administration for War during World War II and at that time vice president and director of Standard Oil Company of New Jersey, was appointed chairman of the operation, and A. C. Long of the Texas Oil Company was named his vice chairman. Headquarters of the entire operation were set up in New York, the location also of the Chase Manhattan Bank, which is generally considered to be the "oil bank" of the U.S. petroleum industry as a whole and of the Standard Oil Company and all its many affiliates in particular.

The Petroleum Administration for Defense also set up a special committee that served as a tool for the dissemination of information. One of its major preoccupations was the equitable supply of aviation gasoline due to the shortages created by the shutdown of the refinery at Abadan and the ensuing two-year boycott on Iranian oil by British and U.S. interests. As a clearing house for such information, the committee could redirect shipments to locations of the most acute shortages. According to the Defense Production Administration,

> The Office of International Trade, being
> an integral part of this foreign petroleum
> committee, functioned valiantly in refus-
> ing to grant export shipments of aviation

gasoline to areas having adequate inventories.  This left the supply group free to
shift intended deliveries to some other
point which was about to run out of avgas.[13]

The supply problem of petroleum, however, "was not
solved on paper. . . .  It was solved by hundreds of separate operation actions in scores of countries."[14]  Under the
voluntary agreement, tankers sailed from the United States
carrying aviation gasoline to airports in the Eastern Hemisphere.  Without those shipments, U.S. planes flying to the
Eastern Hemisphere could not come back.  Various other petroleum products, besides avgas, were also shipped to the
Eastern Hemisphere to keep it from running out of supplies.
    The burden of providing for the shortages created by
their own boycott on Iranian oil did not fall exclusively
upon the shoulders of the domestic corporations in the United States.  Foreign petroleum operations also expanded
production in order to meet the need.  Companies operating
oil wells in Saudi Arabia and Kuwait expanded production,
and the petroleum companies and refineries in the Caribbean
produced to their full capacity.  Thus, consumers who up
until that time had been supplied by the British-owned
Anglo-Iranian Oil Company were now being supplied with oil
from other Middle Eastern countries.  These new supplies
were generally produced by U.S. companies.  The U.S. companies' cooperation with AIOC in boycotting Iranian oil
proved to be, in the final analysis, extremely lucrative
for them.


WESTERN EUROPE IN THE IRANIAN OIL CRISIS

    As seen above, when the boycott on Iranian oil produc-
tion started, the international oil industry was forced
into a series of short-term expedient actions of a temporary nature in order to meet the sudden deficit in world
oil supplies.  This sudden reorganization and redirection
of the world's oil flow resulted in some important long-
term plans of a permanent nature.  To supplement the shortages of oil products from Iran, the European industry had
to draw on its patiently accumulated dollar reserves in order to procure the supply from Western Hemisphere sources.
Moreover, the great distances that some crude oil and products had to be shipped involved heavy additional costs,
partly in foreign currency but mainly dollars.  The imbalance could have caused an erosion of the European balance

of payments had it not been for the pressure put by European governments on their refining industry to increase its
throughput capacity. The effort was so successfully carried out that imports of refined products were actually reduced and exports increased. Thus, the total effect of the
Iranian oil crisis on the major European countries was beneficial.*

The Iranian oil emergency thus inevitably left its mark
on the entire European scene. The OEEC countries depended
on Iran for about 7 million tons a year of oil products and
over 4.5 million tons a year of crude oil, totaling roughly
16 percent of all consumption of oil in that area. "The
loss of this source of oil presented to the petroleum industry a formidable problem, of which a solution was imperative if dislocation of supplies in a number of countries
was to be avoided."[15] European refineries began to operate
almost to the limit of their capacity at a rate not previously foreseen, in an effort to supply markets in Europe
and elsewhere, thus providing a flexibility margin of a potential throughput that had not been anticipated. In order
to accomplish this large-scale production, obsolete units
were put back into operation and bottlenecks were smoothed
out and streamlined by reorganization. Thus, new plans
were made for permanent enlargement of throughput capacity,
and smaller plants found a new lease on life. The AIOC's
agreements with European refineries during the Iranian oil
crisis necessitated a high level of operation, which, in
turn, caused refining operation capacity to increase, particularly in France, Italy, and the United Kingdom. As a
result of these emergency expediencies, European refineries
put into work various expansion plans "to place part of
their output at the disposal of the programs of the international companies, which involved shipments outside Europe."[16] According to the OEEC,

> Western Europe had made an outstanding con-
> tribution, . . . in filling the gap when
> supplies of finished products from Iran
> were shut off from 1951 onwards. European
> refineries have had the capacity, in an
> emergency, to maintain substantial flow of

---

*The only exception was the United Kingdom, which, un-
til alternative arrangements were made, had to shoulder
alone most of the burden of the dollar cost of oil borne
by the AIOC.

exports to non-participating countries, and
it is fortunate not only that expansion had
reached a sufficiently advanced stage at
the time of the crisis, but also that refineries had the margin of flexibility to meet
an exceptional demand. The question arises
whether the outlet in non-participating
countries will be available when new refineries in Australia, India and Aden come on
stream in the comparatively near future, or
if products are again available from Abadan.[17]

After the shutdown at Abadan, the second half of 1951
saw an unprecedented growth in the volume of petroleum products passing through the Suez Canal from north to south,
bringing the total for the year to nearly 2 million tons.
About half of this traffic originated from refineries in
the United Kingdom and France. In 1952 some 4 million tons
of products from European refineries passed through the
canal, of which France accounted for 1.5 million tons, the
United Kingdom for 1.4 million tons, and Italy for about
0.9 million tons.[18]  (See Table 5.)

TABLE 5

East-of-Suez Oil Traffic from European Refineries
(tons)

| Country | 1951 | 1952 |
|---|---|---|
| United Kingdom | 1,000,000 | 1,400,000 |
| France | | 1,500,000 |
| Italy | -- | 900,000 |
| Others | 1,000,000 | 200,000 |
| Total | 2,000,000 | 4,000,000 |

Source: Compiled by the author.

The question that now arose was, How permanent was
this new expansion of European oil products trade to east-
of-the-Suez countries?  It was assumed that when the planned
refineries in Australia and India came on stream in a few
years, while the Abadan refineries would remain immobilized,
there would still remain a market for European products.
If, however, the Abadan refineries were to reopen, expanded

European oil products trade was expected to be able to find
new outlets.  Therefore, no attempt was made to apply temporary measures to the current oil emergency, and more permanent plans were being blueprinted in many European capitals.

## THE IRANIAN OIL CONSORTIUM

The sudden shutdown of the petroleum refinery in Abadan and the stoppage of normal petroleum flow from Iran hit
Europe in the midst of its phenomenal recovery.  The entire
European economy was thus threatened.  However, not only
had the peacetime economy of Europe suffered as a result of
the crisis, but a serious threat also existed to the entire
effectiveness planning of NATO's "shield" posture in Europe.
Abruptly attention became focused on the political implications of this total dependence of Europe on Middle East oil.
The problem of petroleum supply to Western Europe was not
affected merely by a temporary crisis subject to solution
by temporary emergency measures.  Instabilities in the Middle East regimes were the permanent feature of the region,
and further nationalizations of petroleum resources by local governments could be counted on in the future.  Thus
Europe was an easy target for all sorts of industrial and
strategic blackmail.

The political and economic implications of this Western European dependence on unreliable sources of energy supply caused great anxiety in Washington.  Two U.S. administrations became involved in the Iranian controversy.

As soon as the British AIOC was actually ousted from
Iran, the U.S. government offered to mediate.  President
Truman sent his ambassador-at-large, Averell Harriman, to
Iran to work out a solution between the Iranian government
and the British concessionaries.  But Harriman found the
Iranians in no mood for negotiations.  The people were
roused against foreign exploitation of their natural resources, and the government, no matter how it felt about
the economic consequences of this step, could not fail to
see the important political capital gain that it was accru-
ing with the electorate.

Two officials of the Eisenhower Administration became
directly involved in the events that led to the final settlement.  Herbert C. Hoover, Jr.,* a consultant in the

---

*The role of Herbert C. Hoover, Jr., in the mediations
was frowned upon in some quarters because his United Geological Company worked closely with the Standard Oil Company

Department of State since 1953, and the Ambassador to Iran, Loy W. Henderson, acted as liaisons between the British, the Iranians, and the oil companies. As a result of these negotiations, two companies were formed: one, to operate the Abadan refinery, and the other, to operate the oil fields. Both companies, as specified by the agreement, were to be working on behalf of the Iranian government and the National Iranian Oil Company, which were the legitimate owners, from then on, of both enterprises.[19] The settlement between the parties involved became known as the Consortium Agreement. According to George Stocking,

> The parties to the Consortium Agreement are
> Iran and the National Iranian Oil Company
> (parties of the first part) and the several
> consortium members (parties of the second
> part). The National Iranian Oil Company
> (NIOC) is the corporate instrument that the
> government created to own and operate the
> oil industry under Mossadegh's nationaliza-
> tion program. Under the new government NIOC
> became the owner of all producing, refining,
> and auxiliary installations owned by AIOC be-
> fore nationalization and of such other prop-
> erties as the operating companies (created
> under the agreement) may subsequently in-
> stall.[20]

The role of the members of the consortium was distinct-ly different from that of traditional concessionaries. As a group, they did not buy oil nor did they sell it to anyone. According to the agreement they were merely engaged in pro-duction for NIOC. However, again according to Stocking,

> each member of the Consortium set up a trad-
> ing company that individually and indepen-
> dently of the others buys oil at the well-
> head from NIOC at a "stated price" of $12\frac{1}{2}$
> percent of the price that the trading company
> posts at the point of export. The Iranian

---

of California in the United States as well as with Aramco (of which California Standard was a coowner) in Saudi Arabia When the Standard Oil Company of California became eventuall a partner in the forthcoming Iranian oil consortium, Hoover' position as an impartial mediator was questioned.

Oil Exploration and Producing Company then
delivers oil for the account of the trading
company either to the crude oil company
loading ports at the upper end of the Persian Gulf or to the Abadan refinery.  At the
loading port the trading company sells its
crude oil for export usually, but not always
to an affiliate.  The Iranian Refining Company produces the oil that is delivered to
the Abadan refinery and delivers the products to the trading company for sale to its
affiliates or to others.  The trading companies pay the costs which the producing
and refining companies incur in their operations and in addition pay them a fee of one
shilling for each cubic meter of oil that
they receive either for export or for refining.[21]

The new concessionaire, known henceforth as the consortium, was composed of several international oil companies; the British Petroleum Company (formerly AIOC, until
then the sole proprietor of the concession) retained 40
percent of the stock, 14 percent was assigned to the Royal
Dutch-Shell Oil Co., and a 6 percent interest was surrendered to the Compagnie Française des Pétroles.  These
three companies were also partners in the Iraqi concession.[22]  The role of the U.S. government in extending its
good offices in the settlement of the Iranian oil dispute
was acknowledged, and an 8 percent share was allotted to
each of five major American petroleum companies operating
in the Middle East:  Standard Oil Company of New Jersey,
Socony-Vacuum Oil Company, Standard Oil Company of California, Gulf Oil Corporation, and the Texas Company.[23]  The
approval of the new merger, however, aroused public opinion
in the United States.  Louis Schwartz described the effect
of the approval of the consortium to both the Antitrust Subcommittee and the attorney general's Antitrust Committee as
follows:

Objections to the plan from the standpoint
of American antitrust policy are formidable.
Here were huge enterprises already established in various concessions in the Middle
East with more than adequate reserves of
oil.  Some of them had already been officially accused of conspiring to maintain an

artificially high price for the cheap Middle East petroleum. The <u>Wall Street Journal</u>
reported the open secret that one of their
main concerns in entering this pool was to
see to it that Iranian production should
not return to the world market too rapidly
as to hurt the world price.[24]

Acceding to pressures of antitrust groups and domestic
petroleum companies, which feared the increase of economic
power now in the hands of their overseas competitors, the
Department of State insisted that the five U.S. international petroleum companies in the consortium relinquish a
portion of their interest in the Iranian concession to
eight independent U.S. oil companies that had long been
clamoring for a foothold in Middle East oil ventures. Accordingly, each of the big five surrendered 1 percent of
its interest in the concession to the independent companies.
Thus, the following independents were brought into the venture: the Richfield Oil Company, the Hancock Oil Company,
the Signal Oil and Gas Company (which later absorbed Hancock), the Atlantic Oil Company, the Pacific Western Oil
Corporation, the San Jacinto Petroleum Corporation, the
Standard Oil Company (Ohio), and the Tidewater Associated
Oil Company.[25]

## SUMMARY AND CONCLUSIONS

The intensification of the Cold War and the State Department's concern with strong defense alliances increased
the strategic value placed upon petroleum. It was calculated in Washington that a third world war would be fought
predominantly with the aid of petroleum, as the potential
uses of atomic energy were apparently not being made rapidly available. The army, the navy, and the air force would
then be utterly dependent in their operations upon the
petroleum-product-fueled engine.

But defense alliances were only one aspect of this
policy. Since military factors were given much weight in
the determination of all foreign policy in the Cold War
years, these factors were soon introduced into the formulation of economic and foreign trade policies. Starting out
initially with gifts and loans as a form of aid to wooed
nations, the U.S. government soon came to believe that mutual trade was a better-lasting and more binding cement for
alliances in time of peace. "Mutually advantageous trade

relations," President Eisenhower said, "are not only profitable but they are also more binding and more enduring than
costly grants and other forms of aid."[26] Saying that, Eisenhower was, in fact, linking foreign trade with the U.S.
military posture, and his administration proceeded to give
recognition to the military importance of trade with allied
and friendly nations. In particular, the petroleum industry was singled out in view of its great commercial and industrial as well as military potentialities. The Iranian
consortium was one case where the expansion of private U.S.
oil interests abroad was assisted by the Eisenhower Administration for military and political reasons.

Thus, U.S. foreign oil policy in the Middle East took
a new turn. It was observed earlier in this book how in
the 1920s the State Department supported private U.S. investments in Middle East oil as a continuation of the traditional policy of extending protection to U.S. interests held
by private U.S. citizens abroad. This policy was, in the
early 1920s, upheld by the open-door principles and spurred
by the general "oil panic" that seized the country. With
the decrease of the oil scares and the inward direction of
government policies as a result of the Depression, the
State Department generally paid only scant attention to
Middle East oil during the 1930s. World War II constituted
a new turning point in U.S. foreign oil policy inasmuch as
once again the national emergency pointed out clearly the
immense importance of petroleum and all petroleum products
to security. This time, however, the third Roosevelt Administration was not interested in seeing international cartels
take possession of the largest oil resources in the world.
Instead, Roosevelt and his secretary of the Interior desired
to acquire government ownership over Arabian oil fields,
thus effecting a sharp departure from all previous trade
policies of the United States. As already discussed earlier
in this study, such a departure from traditional U.S. concepts was not possible, due to the general commitment of the
entire U.S. system to the principles of private enterprise,
a doctrinal commitment that received material assistance
from the oil lobby in Washington.

Thus, by the time of the Iranian oil crisis (1951-53),
U.S. Middle East oil policy had already undergone three defined phases: (1) traditional diplomatic assistance to private U.S. enterprises abroad (1920s); (2) complete detach-
ment from Middle East oil interests (1930s); and (3) the
emergence of the national interest in Middle East oil and
the unsuccessful attempt made by the U.S. government to subordinate private concerns to the dictates of national security.

The postwar years were marked by increasing U.S. concern over Soviet expansionist schemes. The Truman Doctrine, announcing U.S. assistance to Greece and Turkey, also became a cornerstone of U.S. foreign policy in the years to come. On one occasion, according to George Lenczowski, Truman "described direct Soviet pressures on Turkey and the indirect ones on Greece and linked them to the encircling movement aimed at Iran."[27] Thus, Iran became a focal geographic point of concern to the U.S. containment policy in the Middle East. The consolidation of the Eastern bloc, on one hand, and the diminishing power of Great Britain to hold back Soviet incursions into the Middle East region, on the other hand, again led Eisenhower to abandon the traditional U.S. policy of aloofness from Iranian affairs and to enter, instead, into a firm and binding alliance with that southern neighbor of the Soviet Union.

To what extent was Iranian oil used by U.S. foreign policy as a goal and to what extent was it used as a tool toward an ulterior objective? Inasmuch as the containment of communism in the Middle East was concerned, it is clear that oil could serve as a tool, for U.S. investment capital could get a strong foothold in the economy of Iran and thus become an important agent in the life and politics of that country. On the other hand, Iranian oil had already proved itself to be indispensable to the United States in two cases: (1) for national emergencies such as the limited war in Korea and (2) for the economic and industrial recovery of all OEEC countries of Western Europe. In both cases, then, Iranian oil was a policy goal.

Within these political, economic, and strategic scales of frequently changing balances, the private oil companies seemed to be concerned solely with the purely commercial aspects of the oil trade. Commercially, the companies proved very quickly that they were in a strong position to boycott the Iranian oil industry and that no producing nation could boycott the companies as successfully. The basic truth of the oil imbalance in the world thus became clear during the Iranian oil crisis: Iran depended on Europe for the marketing of its oil. Without Europe, the Iranian oil trade died. But without Iran, the European entrepreneur turned to alternate sources. Therefore, the foundation was thus laid for the then current oil policy axiom that, during oil boycotts, it is the producing nation, rather than the consuming nation, that is bound to be the chief loser.

As a result of the great effort expended in maintaining the flow of Middle East oil going to Western Europe during the Iranian petroleum crisis, government agencies became

alert to the complementary, but inseparable, problem of
sea-borne transportation.  While securing U.S. interest in
Iranian oil production, U.S. officials also became aware
of the interdependence between the well-head and the means
of transportation.  For the Iranian oil crisis was created,
basically, not so much because of oil shortages due to the
boycott on Iranian oil but because of the lack of a sufficient fleet of tankers to carry the available oil and oil
products from their source in the Western Hemisphere to nations in Western Europe and the conflict in the Far East.
The pathetic condition of the U.S. tanker fleets and their
inability to meet the requirements of a limited national
emergency jeopardized the entire defense preparedness of
the country, and, commercially, it handed the U.S. sea-
borne trade in petroleum to foreign tanker fleets.  In
spite of pressures made for construction of a merchant
marine, the budget for such a project depended on approval
by Congress, and this approval was not forthcoming.

# 6

**ANCILLARY
INDUSTRIES**

The oil industry system is an integrated system--that
is to say, it consists of more than one type of economic
system.  It differs from the simple industry system on a
number of counts.  A simple industry system, as distinguished from the integrated industry system, relies on a
single operational chain of actions that are both vertical
and hierarchical.  The simple industry system is concerned
only with direct production of a specified number of related industrial products or services.  In its most simpli-
fied form, it channels marketing and distribution to other
economic actors.  In its more complex form, it does all
that through its own agents, affiliates, or subsidiaries.
The relationship between all parts of the major production
firm, as well as its marketing and distributing arms (when
it owns them), is a pyramidal one.  It is based on concepts
of hierarchy that are called a vertical relationship.
On the other hand, the integrated industry system can
operate only when a number of industrial complexes are coordinated.  The interdependence between each one of these
industries is complete, as none would be able to exist without the others.  This kind of relationship between the various industries gives each one of them a great measure of
equal economic independence within the existing integrated
industry system.  It creates an egalitarian rather than a
hierarchical system, which we call a horizontal relationship.
The oil industry system is such an integrated industry
system, composed of several economic systems, each one operating with a great measure of economic independence within the broader system and, at the same time, maintaining a
constant horizontal relationship with the other economic
branches in that system.

The first and most publicized economic system in the oil industry system is the production system.  Thus far, we have been dealing nearly exclusively with this branch of the oil industry and its relations with various political and governmental problems that influence the foreign policy of the United States.  However, the production of oil at the source would be of no avail were there no tanker fleets to transport it to consuming areas.  The tanker fleets form a separate economic system from the production system.  Oil companies rarely own the tanker fleets needed to fill all their requirements.  They charter a large number of tankers from private owners who may make similar agreements with several other producing corporations.  Nevertheless, these two economic systems, the oil production industry and the tanker service industry, are integrated in the sense that each one of them needs the other in order to continue to function profitably.

The third branch of the integrated oil industry is that of the various types of refineries, including the petro-chemical plants.  There would be no profit for the oil com-panies to produce oil at the well-head, and for the tanker companies to transport this oil, were there no refineries to undertake the preparation of the crude to fulfill con-sumption requirements.  The petrochemical industry, then, is related to the other two industries in the same way as these two industries relate to each other:  It is part of the general oil industry system.  It operates as an indepen-dent economic agent within this system, and its relations with the other two arms of the oil industry are egalitarian and not hierarchical.  Therefore, we call this system a hor-izontal relationship.  All these three arms of the general oil industry, in spite of their individual economic indepen-dence within the general industry system, are completely de-pendent for their continuity on each other and cannot oper-ate profitably except in conjunction with each other.

Finally, we arrive at the distributing and marketing end of the integrated oil industry system.  As in the case of the simple industry system, distribution and marketing may or may not be incorporated in the production or the re-fining branches of the general system, or they may be as-signed to separate agents whose relationship, however, to the general system is nonetheless vertical, as in the sim-ple industry system.

In this chapter we deal exclusively with the two re-maining arms of the integrated oil industry besides produc-tion--namely, the tanker fleets and the petrochemical in-dustry.  The first section of this chapter discusses the poor condition of the U.S. merchant marine and tanker

fleets. The second section describes the striking contrast
posed by the rapid growth of tanker fleets in the OEEC area
as well as such tankers as are registered under flags of
convenience. The third section deals exclusively with the
petrochemical industry in the OEEC area.

## U.S. TANKER FLEETS AND THE MERCHANT MARINE

In historical perspective, U.S. foreign policy, avoiding political entanglements, but otherwise aggressively
pursuing economic gains, heavily depended on means of navigation. These have never been provided in sufficient number either by an adequate merchant marine or by private owners. John Quincy Adams's admonition that two principles
guide U.S. foreign policy, commerce and navigation, has been
honored only in the first part. U.S. trade is carried for
the most part by merchant vessels of other nations. The increased construction of many new ships during World War II,
among which were a large number of ocean-going and coastwise oil tankers, was commonly interpreted as a contribution to the war effort that had no direct bearing on peacetime policies.

The elimination of allied and axis merchant marines as
a result of World War II offered the United States an un-
paralleled but short-lived opportunity to develop U.S. services in all foreign trade routes of the world. But other
nations, taking advantage of the same opportunity, soon surpassed the United States in efforts to fill the vacuum left
by the destruction of those fleets. The United States, however, failed to recognize the challenge posed to its world
position by this large-scale development of foreign merchant marines and neglected to build up its own fleets.

This lack of political and economic foresight soon
caused a crisis situation in all U.S. shipyards. The tanker industry was especially hurt, inasmuch as no more contracts were forthcoming for the building of new merchant
vessels. As a result, work in many of the largest U.S.
shipyards was almost at a standstill (see Table 6). Tens
of thousands of skilled craftsmen were laid off and forced
into other types of employment while formerly thriving
coastal urban centers, which had prospered on shipbuilding,
were now fast losing large numbers of their population.

Not only was the shipbuilding industry hard hit, but
the U.S. merchant marine itself was dying of lack of work.
At that time the merchant marine consisted of antiquated
prewar vessels and the wartime merchant vessels that had

been built under conditions of stress for emergency use and
that were worthless for peacetime purposes. They were un-
suitable either to compete in foreign trade with new foreign-
flag vessels or to form the backbone of a merchant marine
adequate for national security.

TABLE 6

Tanker Construction in U.S. Shipyards
(2,000 gross tons and over)

| Year | Tankers | | Total Construction | |
|------|---------|------------|-------------------|-------------|
| | Number | Gross Tons | Number | Gross Tons |
| 1934 | -- | -- | 2 | 9,544 |
| 1939 | 11 | 119,429 | 28 | 241,052 |
| 1940 | 16 | 148,509 | 53 | 444,727 |
| 1943 | 231 | 2,163,147 | 1,661 | 12,499,873 |
| 1944 | 240 | 2,485,923 | 1,463 | 11,404,404 |
| 1946 | 8 | 80,055 | 88 | 672,554 |
| 1947 | 5 | 22,662 | 50 | 286,473 |

Source: Based on a larger table in U.S. Congress, House
of Representatives, Committee on Merchant Marine and Fisheries, The Merchant Marine Act of 1936 and the Ship Sales Act
of 1946, Hearings, 80th Cong., 2d sess. (Washington, D.C.:
Government Printing Office, 1948).

The depressed conditions in U.S. shipyards and the slow
disintegration of the merchant marine as an economic and
military arm of the nation defeated the purposes of the
Merchant Marine Act of 1936, which had clearly stipulated
the fundamental importance of a merchant marine for the national welfare. "It is necessary for the national defense
and development of its foreign and domestic commerce," reads
Section 101 of the act,

> that the United States shall have merchant
> marine (a) sufficient to carry its domestic
> water-borne commerce and a substantial por-
> tion of the water-borne export and import
> foreign commerce of the United States and
> to provide shipping services on all routes

essential for maintaining the flood of such
domestic and foreign water-borne commerce
at all times, (b) capable of serving as a
naval and military auxiliary in time of war
or national emergency, (c) owned and operated under the United States flag by citizens of the United States insofar as may be
practicable, and (d) composed of the best-
equipped, safest, and most suitable types of
vessels, constructed in the United States
and manned with trained and efficient citizen personnel.  It is hereby declared to be
the policy of the United States to foster
the development and encourage the maintenance of such a merchant marine.[1]

At the same time, when conditions in domestic shipyards were fast becoming desperate, across the ocean,
wounded and formerly vanquished nations, now largely sub-
sisting on U.S. relief, were engaged in shipbuilding programs.  Great Britain alone was constructing more than half
of the world's postwar merchant shipping (see Table 7).*
Sweden, France, Holland, Italy, and the British Dominions
exceeded U.S. total tonnage of construction.  All these
foreign shipyards were busy and had ample work for several
years.  Total U.S. shipping construction in the postwar
years exceeded only that undertaken by Denmark and Spain.
In a matter of two years, the United States slipped from
first place in world shipbuilding to seventh place.

---

*In 1945 the United States still built six times the
tonnage of Great Britain.  In 1947 Great Britain exceeded
the United States seven times.  On January 30, 1946, there
were under construction in Great Britain 2,062,949 gross
tons of merchant marine vessels, which accounted for 53.6
percent of the world total of 3,846,656 gross tons under
construction at that time.  At the same time, the United
States was constructing only 183,236 gross tons, a ratio
of 9 percent of the British construction.  British shipyards at that time were building to capacity and had a
backlog of orders sufficient to engage them to capacity
through 1950.  (See U.S. Congress, House, Committee on
Merchant Marine and Fisheries, The Merchant Marine Act of
1936 and the Ship Sales Act of 1946, Hearings, 80th Cong.,
2d sess. [Washington, D.C.:  Government Printing Office,
1948].)

TABLE 7

Merchant Shipping, by Country, January 1948
(1,600 gross tons and over)

| Country | Tankers | | Total Merchant Vessel Construction | |
|---|---|---|---|---|
| | Number | Gross Tons | Number | Gross Tons |
| Great Britain | 137 | 1,737,402 | 545 | 4,250,404 |
| Belgium | 5 | 78,475 | 20 | 167,129 |
| Canada | -- | -- | 26 | 136,278 |
| Denmark | 6 | 76,800 | 57 | 322,760 |
| France | 9 | 140,163 | 61 | 448,340 |
| Holland | 7 | 84,700 | 72 | 378,431 |
| Italy | 2 | 9,500 | 46 | 282,060 |
| Norway | 3 | 31,500 | 62 | 194,095 |
| Spain | 2 | 19,350 | 20 | 115,700 |
| Sweden | 72 | 693,475 | 206 | 1,284,570 |
| U.S. | 19 | 289,990 | 43 | 408,318 |

Source: Based on a larger table in U.S. Congress, House, Merchant Marine Act of 1936 and Ship Sales Act of 1946, Hearings, 80th Cong., 2d sess. (Washington, D.C.: Government Printing Office, 1948), p. 121.

The comparison between work done in domestic shipyards and the high rate of production in foreign countries caused a good number of U.S. shipbuilders and congressmen to wonder how these foreign nations, despite their desperate economic conditions and their high degree of reliance upon the United States for economic assistance, could afford such heavy shipbuilding programs, while the United States could not. Neither could they understand any better why the 16 countries participating in the Marshall Plan could plan a ship construction program that would increase their fleet tonnage to a point where it would surpass the prewar tonnage, so that by 1951, it was surmised, it threatened to eliminate completely the U.S. merchant marine from any further participation in the reconstruction of Europe.

The low status of the merchant marine and the shipbuilding industry occupied President Truman's attention when he appointed an advisory committee to study and suggest measures to correct the situation. The committee very quickly reached three conclusions--namely, that a modern, efficient merchant fleet and a progressive shipbuilding

industry were "indispensable to national security"; that
they were also "sufficiently essential adjuncts to our
peacetime economy" to justify any government financial aid
necessary in order to maintain it; and that the United
States ought to lead international shipping, as it was expected to lead in all other world affairs.[2]

The advisory committee transmitted its report and
recommendations to the president on November 1, 1947.  It
advised the president that the merchant marine and the shipbuilding and ship-repairing industries were interdependent
and that the modernization and the efficiency of both were
indispensable to national security and that this entire industrial capacity must be able to be expanded instantly in
the event of emergency to meet wartime requirements but
that, in current conditions, it could not respond instantly
as required in emergency unless the industry as a whole was
sound and healthy at all times.  The committee further advised the president that any allocations for the construction of new ships should remain within the boundaries of
private ownership.  It concluded that

> the best prospect for a strong progressive
> merchant marine would be realized in one
> that is privately owned and privately op-
> erated on a remunerative basis in commer-
> cial services, receiving assistance from
> the Government to the extent necessary to
> put it on a fair competitive basis in its
> various foreign trade routes and to pro-
> tect it from discriminatory practices in
> domestic trade.[3]

The recommendations submitted by the president's advisory committee to direct government financial aid toward
the shipbuilding industry, the gloomy forecasts by the Maritime Commission on the sad state of the merchant marine,
and the warnings issued by the Petroleum Administration for
Defense as to the grave consequences of this industrial ebb
to the entire security complex of the nation did not move
Congress to allocate sufficient funds in order to revive
the shipbuilding and ship-repairing industries and maintain
a modern and efficient merchant fleet.  Lack of knowledge
as to the future regarding unforeseen changes of supply and
demand automatically and unavoidably introduced an area of
underestimation of transportation requirements.[4]  This un-
certainty was not confined to matters of civilian shipments
alone.  The Military Sea Transportation Service, whose

nucleus fleet provided the degree of flexibility demanded
by military operations, was dependent for its efficiency on
a careful coordination of commercial capabilities with military requirements, both of which constantly fluctuated.
If, however, privately owned U.S. flag commercial vessels
were not available, shipments for military requirements
would then be delayed and the nucleus fleet would not be
able to provide the necessary capabilities for sudden exigencies.[5]

Since the threat of European competition to the U.S.
merchant marine very quickly became a reality in the postwar years, it was necessary to find means by which to assist this branch of the water transportation of the United
States from dying away. This acute condition was sharply
pointed out by the Korean War and the Iranian oil crisis.
At the commencement of the conflict in the Far East, military logistics depended heavily on the nucleus fleet of
tankers belonging to the Military Sea Transportation Service. Alone, however, the nucleus fleet was unable to satisfy the bulk petroleum transportation requirements. The
supplementary requirements of the war were carried by privately owned tankers and through the voluntary tanker pool
that started its operations in December 1950.

The depression in the tanker industry and its relation
to the national economy became once again apparent during
the Iranian oil crisis of 1951-53, when a large number of
tankers was suddenly required in order to divert shipments
of petroleum and petroleum products into different routes.
That dispute between Great Britain and the government of
Iran made the U.S. people realize the extreme dependence
of their country's commerce on foreign means of transportation. There was a growing awareness in Washington that un-
stable conditions in foreign countries might thus jeopardize
the smooth sailing of U.S. commerce in world trade routes.
The Merchant Marine Act of 1936, which still existed as the
basis of the current law, no longer seemed able to cope with
problems that had emerged as a result of new postwar conditions. It was felt that it was necessary to amend the act
and to propose new legislation. The 1950s saw many such
attempts, most of which were unsuccessful.

Eventually a subsidy program did emerge for the construction of sea-going tankers in domestic shipyards. This
U.S. shipbuilding policy in nearly its entirety was intended for such vessels as would sail under registration in
the United States. These construction programs were heavily dependent on government orders and subsidies, as U.S.
prices made an unsupported shipping industry a losing

proposition.  Ships built in the United States cost twice
as much as similar ships built in European shipyards.[6]  But
for the lack of other programs, the international carrying
trade in petroleum was rapidly slipping out of U.S. hands.

In comparison with U.S. programs, Japan's shipyards
had constructed 663,000 gross tons of sea-going vessels between March 1953 and March 1954 alone, a postwar record exceeding the previous year's Japanese shipping construction
program by 122,000 gross tons (Table 8).

TABLE 8

Japan's Shipbuilding Construction

| Period | Gross Tons | Foreign Orders |
|---|---|---|
| March 1952-March 1953 | 541,000 | 165,000 |
| March 1953-March 1954 | 663,000 | 257,000 |

Source:  Compiled by the author.

It seems that while all concerned had agreed that the
merchant marine should be assisted, there was no agreement
as to what form the assistance should take.  Such devices
as direct subsidies, cargo preference,* manipulations or
prohibitions of foreign currencies, reduced port, consular,
and documentation charges, and liberal tax benefits were
considered separately and in combination, but no sizable
positive policy came into being.

One of the arguments against the restriction of cargo
preference was that the U.S. merchant marine must be sacrificed in order to close the international dollar gap.  This
argument was the lead attack upon the provisions of U.S.
laws that stated that American bottoms must carry at least
50 percent of all military or economic aid cargoes as well
as cargoes financed by the U.S. government.  Under these
laws, within a matter of five years, the international dollar gap was reduced from $11,478 million in 1947 to $4,709
million in 1953 (Table 9).[7]

---

*The maritime industry advocated that a substantial
portion of all overseas cargo originating as a result of
the expenditure of public funds should be transported in
U.S. flagships.

TABLE 9

Narrowing Dollar Gap in Goods and Services
(millions of dollars)

| Year | Export | Import | Gap |
|------|--------|--------|-----|
| 1947 | 19,796 | 8,318 | 11,478 |
| 1950 | 14,425 | 12,128 | 2,297 |
| 1951 | imbalance due to oil crisis | | 5,164 |
| 1952 | 20,649 | 15,794 | 4,855 |
| 1953 | 16,437 | 11,904 | 4,709 |

Note: In the year 1947, the peak of the dollar gap between the U.S. exports and imports of goods and services was reached.

Source: U.S. Congress, Senate, Committee on Interstate and Foreign Commerce, Cargo Preference Bill, Hearings Before a Subcommittee on Water Transportation, 83d Cong., 2d sess. (Washington, D.C.: Government Printing Office, 1954), pp. 127-29.


OEEC TANKER INDUSTRY

While U.S. shipyards were gathering dust and the nation's merchant marine was dying for lack of orders, shipbuilding in foreign countries was increasing rapidly. By January 1953 the heavy postwar building program in Western Europe had reached its current apex with outstanding orders amounting to 14.5 million deadweight tons.[8] In 1954 tankers constituted 57.3 percent of the world total tonnage under construction.[9] Total tanker tonnage owned by member countries of the OEEC increased from 13.3 million tons in 1954 to 14.7 million tons in 1955, and their percentage of the world tanker fleet rose from 54 to 55.4 percent. Correspondingly, the percentage of tankers in the total fleets of these countries increased from 28.7 percent in 1954 to over 30 percent in 1955[10] (Table 10).

At the time of the Suez Canal crisis (1956-57), tonnage under construction or on order in the world amounted to 34.5 million tons, nearly twice the amount of existing world tanker tonnage. Almost 70 percent of total world construction took place in European shipyards, which were fully employed during this period.[11]

TABLE 10

Principal Tanker Fleets
(gross tons)

|  | June 30, 1955 |
|---|---|
| OEEC Countries |  |
| Norway | 4,173,000 |
| United Kingdom | 5,261,000 |
| Netherlands | 893,000 |
| France | 1,152,000 |
| Sweden | 902,000 |
| Italy | 1,210,000 |
| Denmark | 504,000 |
|  |  |
| Other Countries |  |
| Liberia | 2,356,000 |
| Japan | 743,000 |
| Panama | 2,134,000 |
| United States | 4,321,000 |

Source: Organization for European Economic Cooperation, Maritime Transportation (Paris, 1955).

Of these world tanker fleets only 5 percent were then over 25 years old and 3 percent were between 20 and 25 years old. It is interesting to note that while the U.S. merchant marine consisted of old and decaying tankers, 83 percent of the tanker tonnage under the Liberian flag, a large percentage of which belonged to U.S. owners, was under 10 years old as compared with 52 percent of the rest of the world.[12]

The practice of registering vessels under flags of convenience raised serious international shipping problems. National policies, as in the United States, led to flag discrimination, while in other countries they caused a slow turnabout of shipping in some ports. The effect of U.S. flag discrimination policy on the maritime nation members of the OEEC was extremely negative. "Flag discrimination has a boomerang effect on those who practice it," the OEEC Maritime Transport Committee's report warned. "If maritime countries cannot sell their shipping services freely on a competitive basis their ability to buy abroad is diminished."[13] Further, it advised that "thus the non-maritime countries also have a great interest in preserving the principle of freedom of shipping services."[14]

This opinion was in accord with indicated world trends, which consistently reported a steady growth of tonnage registered in Panama, Honduras, Liberia, and Costa Rica. Several advantages accrued from such practices. The earnings of ships registered under these flags were not subject to taxation in the country of registration. Since such registration was completely open to all nationalities, the ships under these flags were owned almost exclusively by shipowners of other nations. Such a registration involved merely the payment of a nominal fee. The only registration fee under the Liberian flag cost $120 per ton and thereafter an annual tonnage tax of 10 cents per ton. Honduras and Costa Rica imposed similar minimal charges. These fees represented a substantial contribution to the relatively small economies of these countries.

The small fee paid to national exchequers for registration under those foreign flags enabled shipowners to accumulate reserves that they set aside toward replacement of their tonnage, to sustain trade in times of depression, and to build new tonnage with far less regard to cost than those shipowners who registered their ships under their own national flag could afford.

Since World War II, however, the principle of shipping freedom was being increasingly threatened by administrative measures taken by governments desiring to protect their own national fleets or even to create such fleets. The OEEC policy on liberalization of trade in the field of maritime transport came into conflict with U.S. legislative measures for exchange control, national flag preferential treatment, taxation, and trade licensing. Many attempts were made to reconcile U.S. and Western European policies by formulating a new joint approach to the problem. Over the years, these annual meetings between U.S. commissions and representatives of the OEEC helped to clarify the diverse economic interests of the parties but produced very little positive contribution toward a mutual solution.

It is rather significant that in Europe, far more than in the United States, shipping was seen as being more than merely an arm of the national defense and the security of international trade routes. European governments attached great importance to the role of their shipping earnings in relation to their balance of payments. In 1953, in both Denmark and Sweden, the income from shipping amounted to 15 percent of the total income from foreign trade and the credit balance of payments on shipping account contributed largely toward those countries' total credit balances. In the same year, in Norway, shipping receipts amounted to almost 45 percent of the total receipts from foreign trade, and the net balance of shipping amounted to $176 million, providing a very large contribution to Norway's total balance of payments.[15] Again in the same year, in Holland, transport represented approximately 30 percent of the net balance on current account, out of which shipping contributed a very substantial portion.[16] As U.S. interest in shipping was motivated less by considerations of international credits and balance of payments accounts and more by national security requirements, repeated European proposals for liberalization of sea-borne trade services passed largely unheeded.

The same type of reasoning was responsible for the fact that while European merchant fleets of all types were growing by leaps and bounds, the U.S. merchant marine was fading away. The severe consequences of this policy for the United States became apparent during the Suez Canal petroleum crisis of 1956-57, when, under conditions of international emergency, U.S. oil companies cooperating with U.S. government agencies were forced to look elsewhere for pooling means of transportation to carry their petroleum. Tankers hoisting flags of all maritime nations carried Middle East and Caribbean petroleum, produced by U.S. investment capital, to refineries in Western Europe and Japan. For a period of time, as a result of this shortage in U.S. seaborne tanker fleets, the world petroleum industry became truly internationally interdependent. In the aftermath of the Suez Canal petroleum crisis, it achieved all the characteristics of international industry, inasmuch as production, transportation, refining, and marketing of many different nations became strongly interwoven.


OEEC REFINERIES AND PETROCHEMICAL INDUSTRIES

As could be anticipated, the effort exerted by Western Europe to meet its temporary fuel shortages during the Iranian oil crisis bore lasting and beneficial results for its economy. The expansion of the European refinery industry, which took place under the stress of the shutdown of the refinery in Abadan, supplied many new jobs and thus increased the average income in each country. In turn, the standard of living, which is so closely related to individual income as well as to the consumption of energy, was also raised. Thus, "the average income in each country and the standard of living" were seen as being "closely related to the consumption of energy per capita," and productivity soon came to depend "on the number of horse power used per worker."[17]

Since the shutdown at Abadan had caused such a great expansion of the European economy, its immediate effects were soon noticed in a domestic revolution that broadened the employment of oil products in housework and raised the level of personal comforts and health. Moreover, it had a tremendous effect on international communication by railway, ship, car, and airplane. The ease with which Europeans now traveled across national borders intensified international communication and contributed toward regional unification. It brought the countries closer to each other by promoting international trade. Although all these fundamental changes in the economic structure of the OEEC countries were brought about as a result of what was expected to be only a temporary interruption in the normal flow of petroleum supply from its traditional source, Europeans were nevertheless confidently charting long-range plans.

Under the impact of the Iranian shutdown, Europe became prepared to step outside of its own continental boundaries at the right time, when many emerging nations would become serious new customers on the international market for the first time.* The rising standard of living not only in Europe alone increased the demands on refinery productivity, and European manufacturers quickly adapted their blueprints to fit a worldwide market. The participation of European manufacturers in international trade not only helped expand the economy of the OEEC area but also had important political effects. Many Europeans believed that they ought to bolster their competitive position in international trade in order to resume Europe's position in the world. The OEEC, which originally was set up for the distribution of the Marshall Plan and through which the energy industries of member countries were also helped, expanded its role now as an independent agent of European economic and political aspirations. The growing community of interests of the Western European nations provided a strong argument for the closer coordination of their trade policies. The need for open channels of information was soon complemented by a series of agreements articulating the coordination of their energy policies. The increase of cooperation in the field of energy included plans for exchange of energy; supply of capital needed for the production of energy;

---

*The same consideration of dollar deficits that prompted Western Europe, during the postwar years, to divert its imports of petroleum and petroleum products from U.S. sources also motivated the purchasing policies of the emerging nations.

exchange of manpower for the same purpose; exchange of information and experience; pooling of research and development programs; and a machinery for the coordination of all these programs.[18]

Toward the end of 1953, consideration was already given to the liberalization of trade in petrochemicals between the OEEC countries in addition to trade liberalization already effected by Austria, France, and Germany in other products.

By the end of 1953 the rate of increase in chemical production had exceeded the rate of increases in total industrial production of the OEEC countries (Table 11).[19] The rapid expansion of the petrochemical industry encouraged commensurate increase of investments (Table 12).

TABLE 11

Increase in Chemical Production, 1950-53
(percent)

| Country | Increase |
|---|---|
| Italy | 57 |
| Germany | 43 |
| Belgium | 22 |
| United Kingdom | 20 |
| Norway | 15 |

Source: Based on information in Organization for European Economic Cooperation, The Chemical Industry in Europe (Paris, 1953), p. 37.

In 1956, prior to the Suez Canal petroleum crisis, further planned increases envisioned the investment of $452 million in the petrochemical industries of these countries. The largest single expansion, costing $130 million, was planned in France, thus quadrupling the size of the petrochemical industry of that country. A similar investment of $127 million was being planned by the United Kingdom for the same period of time, and Germany had prepared a blueprint for an investment of $90 million for that year.[20]

The success of these various plans depended directly on the constant availability of an adequate supply of crude oil required to meet the growing needs of Europe and overseas territories. Most of Europe's petroleum supply was purchased from Middle East sources, and the abrupt interruption of their normal flow as a result of the Suez Canal crisis of 1956-57 not only jeopardized much of the planned economic expansion of the OEEC nations but also threatened to depress the entire standard of living of the continent, which had been raised so successfully since 1946.

TABLE 12

Investments in Petrochemical Plants of France, Germany, Italy, Netherlands, and U.K.

| Period | Investments (dollars) | Production (metric tons) |
|---|---|---|
| to July 1, 1953 | $240,000,000 | 208,000 |
| Dec. 31, 1954 | 290,000,000 | 305,000 |
| Dec. 31, 1955 | 314,000,000 | 397,000 |

Source: Organization for European Economic Cooperation, The Chemical Industry in Europe (Paris, 1956), pp. 87, 90-91.


SUMMARY AND CONCLUSIONS

U.S. foreign oil policy was assigned to the executive jurisdiction of the Department of the Interior. Viewed through Interior's prism, foreign oil was seen primarily as an adjunct to domestic supply-and-demand imperatives, subject to rules governing national conservation policies, the protection of the U.S. oil industry, and antitrust legislation. The Oil and Gas Office of the Department of the Interior has never been required to give attention to problems of oil transportation in the Far and Middle East. It was not expected to give answers to questions arising from the competitive European market in petrochemicals or to provide solutions to deficits in U.S. balance of payments. Therefore, the broader view of the oil industry as an integrated industry composed of various complementing and ancillary components has never been developed and brought to its logical conclusion. Foreign oil policy has thus been reduced to a unidimensional concept consisting of the production of oil at the well-head alone. As many other factors outside the scope of the Department of the Interior but nevertheless demanding legitimate attention had to be recognized as well, the Office of Oil and Gas was never able to formulate a U.S. foreign oil policy.

Among these factors, vying for recognition, were the Defense Department's world strategies, the navy's logistical system, the merchant marine as well as private shipbuilders and shipowners, the international banking concerns, the Justice Department, and many others.

Congress, attempting to clarify a national foreign oil policy, has not fared any better than the Oil and Gas Office under all these multiple drives and pressures. Oddly enough, no one in the Senate or House of Representatives has ever questioned the validity of the Department of the Interior's initial approach to the concept of the oil industry as a simple industry system: The Congress, just as the executive, has always viewed the question of foreign oil policy as one of production alone and isolated it artificially from the other two major components of the integrated industry, with the result that Congress has never been able to recommend a U.S. foreign oil policy.

If the oil industry is viewed as a simple industry, it is not surprising that Secretary Krug, in 1948, finally relinquished the task of formulating a foreign oil policy to the National Petroleum Council (NPC), an independent organization representing the private oil sector. It was this private organization that, in the end, produced in effect, the definitive U.S. foreign oil policy, a draft plan that still constitutes the foundation of present-day policies. Having been conceptualized by oil producers, this formula continues to treat oil as a problem of production alone.

This chapter has analyzed the oil industry system and its different industrial components. It has proved that in no case can a sound oil policy be based on the concept of production alone. The conclusion that emerges from this chapter, then, is that the government, in forming a new national oil policy, must correlate at the same time new national merchant marine and shipbuilding policies and, similarly, a new policy regarding foreign trade in petroleum products. Such a coordination between all ancillaries of the same integrated industry must be placed at the very foundation of a sound policy, and the lack of such a coordinated plan might very well be responsible for the failure of any proposed policy.

Thus far, we have attempted to trace the great variety of private and public interests that have converged to make U.S. Middle East oil policy. We have analyzed the outlook and objectives of each one of these groups in relation to Middle East oil and defined the <u>locus</u> of gravity for decision-making at each period of time and under each particular set of conditions.

Each one of these groups had a legitimate right under the law to press for a policy that would best fit those ends. Thus, the Oil and Gas Office, the Justice Department, the navy, the independent producers, the local refiners, the tanker owners, and a large number of other concerns competed against each other in order to direct public policy. They battled it out in congressional hearings, in executive commissions, in the press, and in special study reports. It has been seen, however, that these efforts were constantly subject to examination under the broader guideline of existing legislation as well as the main policy issues of the United States. Thus, legislation in regard to oil conservation policy, antitrust law or the Defense Production Act, and U.S. foreign policy issues in regard to the Soviet Union, the Middle East region, or the NATO alliance have constantly imposed their frames of reference upon any discussion of oil policies.

Emergency situations such as World War II, the Korean War, and the Iranian oil crisis helped to clarify the scale of U.S. priorities in regard to Middle East goals. No crisis in the past, however, so much brought together all public and private groups concerned with oil as the Suez Canal oil crisis of 1956-57. This crisis is the classic case study on the U.S. power structure in relation to Middle East oil,

and the manner of its resolution points out sharply the dominant imperatives of the U.S. scale of national priorities.

In order not to be sidetracked into other considerations
of the Suez Canal crisis that have no significant bearing
upon the question of Middle East oil per se, this chapter
omits any discussion of problems not directly related to U.S.-
European oil operations during the Suez Canal crisis. Thus,
no mention is made of the military operations, the role of
the State of Israel, or the reaction of the Arab nations. If
these omissions may seem arbitrary, one can only note that in
this case, where a detailed search is made into the organization of the oil power structure, the boundaries of the subject matter must be restricted solely to relevant areas.

The first section of this chapter discusses at some
length Britain's traditional and contemporary concepts of its
relation with the Middle East, the resultant fear of sole dependence upon a single trade route via the Suez Canal, the
attitude of the French government, and some problems arising
out of U.S. and Soviet policies. The second section of this
chapter discusses the situation of acute oil shortages that
occurred in Europe as a result of the stoppage of all traffic
through the Suez Canal, and the third section describes the
sources of assistance offered to Europe in order to meet the
continent's energy requirements. Most of the assistance in
crude oil was offered from U.S. and Caribbean sources. This
assistance, however, was subject to important domestic legislative limitations: the conservation policy of the United
States, described in the fourth section; the Defense Production Act, discussed in the fifth section, and the antitrust
law provisions, referred to on various occasions throughout
the chapter. The sixth section of this chapter analyzes the
Voluntary Agreement Relating to Foreign Petroleum Supply,
which was made within the concept of Article 708 of the Defense Production Act. The machinery for the operation of the
Voluntary Agreement became known as the Middle East Emergency
Committee (MEEC). The workings and operations of MEEC, which
reflected the various relationships among the different components of the U.S. power structure in relation to Middle
East oil as well as U.S. foreign policy in that region, are
described in the seventh section.

EUROPE'S CONCERN WITH THE NATIONALIZATION
OF THE SUEZ CANAL

In July 1956, the Western alliance was abruptly faced
with a transportation problem of the utmost magnitude. The

Suez Canal, serving as the waterway link between the source
of energy supply in the Middle East and the European markets,
was suddenly nationalized. Several noteworthy events had
preceded it.

In the first place, Egypt was cultivating its ties with
communist nations. China became the largest buyer of Egyptian cotton, while overall trade with communist nations increased in 1955 to 35 percent of total Egyptian foreign commerce. Arms followed trade. In September 1955, Egypt made
its first large arms deal with Czechoslovakia. Gamel Abdul
Nasser announced that he had concluded with Czechoslovakia
an agreement for the purchase of arms and military equipment.
It was subsequently officially disclosed in London that this
arms deal gave Egypt at least £150 million (about $500 mil-
lion) worth of military equipment.[1] The announcement of
this agreement made a great impact on Great Britain and the
United States. In both countries, it was widely believed
that this deal would inevitably upset the balance of military force in the Middle East, with a resultant marked advantage to the increasing prestige of the USSR in that region. Only a short time before, the Egyptian government had
rejected a proposal for a military pact with other Middle
Eastern and European nations. On July 19, then, the United
States informed the Egyptian government that it was "not
feasible in present circumstances" to take part in the financing of the Aswan Dam. Similar statements followed from
the British government and the World Bank. The sums involved
were a $55 million contribution by the United States and a
$15 million contribution by Great Britain. The World Bank's
proposed loan of $200 million was contingent upon these two
offers and was now automatically withdrawn.[2]

At that stage, the Soviet Union offered to help finance
the construction of the dam and Nasser decided then to supplement the Soviet offer from the profits of the Suez Canal.
The decision taken by Nasser on July 21 to nationalize the
Suez Canal was ratified by the Revolutionary Council on
July 23 and announced to a crowd of 200,000 at Alexandria on
July 26, a date that also coincided with the fifth anniversary of Egypt's revolution and the fourth anniversary of the
abdication of King Farouk. In that historic speech Nasser
referred to a "conspiracy between Britain, America and the
World Bank . . . to induce . . . conditions restricting
Egypt's independence."[3] On that same day, the Government
of Egypt promulgated Law No. 285 of 1956, for the nationalization of all assets in Egypt of the Universal Maritime
Canal Company. The new Egyptian successor company was
granted a concession to operate the canal as an Egyptian

public utility enterprise, and the Suez zone was declared by law to be an integrated part of Egypt.[4]

The nationalization of the Suez Canal was widely viewed at the time as a spontaneous reaction by an irate leader to the U.S. withdrawal from the Aswan Dam project, and Nasser's speech of July 26 was equally said to be an impulsive outburst of irrational behavior. Thus, it might be assumed that Nasser, presumably acting irrationally, was not calculating at the time of the nationalization of the canal the possible military consequences that he was provoking. But not all observers agreed with this view. To at least one commentator it seemed that Nasser

> calculated that Britain would not be able to invade Egypt for four months because the bulk of her military equipment in the Middle East was locked up in the Canal Zone base under Egyptian guards. He knew that France had its army tied down in Algeria and assumed that Russia would have to support Egypt in order to preserve its newly-won prestige in the Arab world.[5]

Nationalization of the canal seemed to pose an immediate threat to the security of the sea-borne commerce of the Western European nations, whose economies depended to a very large degree on Middle East petroleum and petroleum products. The impact of the nationalization on the European nations was so great that it challenged long-standing economic concepts that were the bases of the entire mercantilist system of the West. Since the disintegration of the feudal system, the maritime nations of Europe had formulated their military and foreign policies upon the very fundamental principle of the protection of their trade routes. Two factors--the extension of outlying trade-posts-turned-into-empires, which had, originally, fitted neatly into the mercantilist system, and the principle of freedom of the seas, which had evolved under the laissez-faire economic system--had both emerged as a result of these basic national maritime policies. Now, the nationalization and state monopolization of a most important international waterway by a single country challenged the entire Western politico-economic conceptual structure of the past several hundred years. The act of nationalization by an impulsive and unfriendly Arab leader seemed not only to imperil a short-cut trade route to the Persian Gulf and the Far East but also to threaten the entire base of the international economic system. It was felt by many European

governments that no maritime nation could afford to allow
Egypt to expropriate the canal and that only a firm stand,
supported by all maritime nations, could remedy this breach
of international equity.

The governments of Great Britain and France became particularly concerned with the possible ramifications of this
action of nationalization.  Viewing Egypt's confidence as a
result of its recent ties with the Soviet bloc, neither Prime
Minister Sir Anthony Eden nor Premier Guy Mollet believed
that economic and political pressures alone could attain
their objective.  Since both heads of state were determined
to safeguard the vital interests of their nations, the only
remaining alternative was a military action.  The preservation of their access to Middle East oil thus became _prima
facie_ cause for their attack on Egypt.

It was estimated in London that "the United Kingdom had
reserves of oil which would last for six weeks, and that the
other countries of Western Europe owned comparatively smaller
stocks.  The continuing supply of fuel, which was a vital
source of power to the economy of Britain, was now subject
to Colonel Nasser's whim."[6]  Great Britain imported oil from
Iraq, which flowed west by way of pipelines to the Eastern
Mediterranean.  However, much of the Iranian output of oil
and its products, as well as the production by British Petroleum Company in the Persian Gulf and in Burma, came through
the canal.  This trade amounted to more than half of Britain's annual imports of oil, which was at the mercy of
Colonel Nasser.  In 1954 and, again, in 1955, British shipping tonnage crossing the Suez Canal amounted to 33 million
tons, of which about one half consisted of British oil tankers carrying approximately 20.5 million tons of oil and
representing 65 percent of general British consumption.  It
was expected that this rate-of-consumption pattern would
double by 1985 and that it could only be satisfied by Middle
East sources.  South American and Caribbean oil was largely
prohibitive as sources of supplementary supply for two reasons:  (1) the United States reserved these resources in
order to supply its then small demand for oil imports, to
assist in the U.S. oil conservation policy scheme, and to
retain available additional oil supplies for defense purposes in case of a national emergency; (2) at the same time,
South American and Caribbean oil was too expensively priced
in hard dollar currency, which Great Britain and France
could not afford to draw out of their accumulated balance-
of-payments credits.[7]

In view of the worsening of the political atmosphere
between Egypt and England, it seemed that at any time the

Egyptian government might decide to interfere with British passage rights through the canal and to persuade Syria to cut the pipelines belonging to the Iraq Petroleum Company, which passed through its territory. These reasonable possibilities had to be gauged against the alternative of bringing in oil from the Middle East by the long haul of tanker transportation around the Cape of Good Hope.[8]

> It was not enough for Egypt to affirm that she
> had no intention of changing her attitude to
> passage through the Canal; arbitrarily, she
> had already done so: Israeli vessels were de-
> barred from using it. Soon, other nations who
> were not politically in accord with Egypt might
> be denied passage, whilst the Canal dues could
> be raised selectively according to the whims
> of the Egyptian Government.[9]

The fear of what the unpredictable Colonel Nasser might do spread quickly throughout the British capital. A lead article in the Times of London of August 15, 1956, warned the Conservative government of Sir Anthony Eden that if Nasser were "allowed to get away with his coup all the British and other Western interests in the Middle East would crumple." According to this editorial, the imperative assumptions of Western policy ought to be (1) that the freedom of passage through the Suez Canal in peace or war be maintained; (2) that this freedom could be assured if the canal were in friendly and trustworthy hands; (3) that Nasser was neither friendly nor trustworthy; (4) that Middle East oil was at the foundation of European industry and Western security and, therefore, its safe conduct was a precondition to Western Europe's survival; and (5) that the unilateral abrogation of international treaties would undermine world stability if allowed to pass with impunity.

The Times' view of Nasser's intentions was reflected in that of the prime minister. "If Nasser is allowed to defy the eighteen nations," Sir Anthony Eden was convinced that it would merely be "a matter of months before revolution breaks out in the oil-bearing countries and the West is wholly deprived of Middle East oil."[10]

It was not the intention of Eden to let his nation perish by degrees for lack of fuel. Considerations other than those of the scarcity of fuel also motivated the French premier. "If Egypt were allowed to succeed in grabbing the canal," Guy Mollet declared, "the Algerian nationalists would take fresh heart. They would also look to Egypt for backing, which they would certainly receive, both in arms and clamour."[11]

To all the Western European pleas for cooperation, the United States responded only by freezing the Egyptian dollar balances. At the same time, Secretary of State John Foster Dulles began to question the legal grounds of Europe's grievances and refused to support the Franco-British resolution in the United Nations Security Council. However, U.S. and European economic pressures failed to produce the desired results, as Egypt was getting help from nations of the Soviet bloc. According to A. J. Barker, "This confirmed any scepticism about the effect of economic sanctions. It strengthened the need for speed in our diplomacy and in any possible military sequel."[12]

The Soviet role in the canal crisis was not to be minimized. Since the Federal Republic of Germany had joined the NATO Alliance, the Soviet Union became active on many different fronts simultaneously. "To begin with," Robert Hunter has written, "there was no longer any need for the Russians to maintain the fiction that Eastern bloc relations were less formalized than those of the West," and, therefore, "within months of West Germany's entry into NATO, the Soviet Union constituted a military pact of its own in the East by a treaty signed in Warsaw on May 14, 1955."[13]

Under these Cold War conditions, the opportunity to strangle a major Western trade route at its most sensitive point on the Suez Canal could not have escaped the globe-embracing eye of the Soviet leadership. The great significance attached to Egypt and the Middle East as a communication highway between East and West had been a fundamental principle of international relations since Marco Polo and Vasco da Gama. It was this principle that had brought France to Egypt in the first place. When the directoire had appointed Napoleon Bonaparte to be commander-in-chief of the armies of the Republic, it had instructed him to subdue England by invading the British Isles. However, knowing that the invasion of England through the channel was an objective far beyond the capabilities of the French Army under his command and further knowing that even if successful the British Navy would still be at large while every last French soldier would be stationed in England in order to enforce the occupation, Napoleon, after assuming command of the French armies, had deviated from the instructions issued to him by the directoire. In his military genius, and fully aware of the major role in battle of the indirect approach, he had decided to bring England to her knees by striking at her stomach. Instead of landing with his armies on the banks

of the Thames, he had arrived with large expeditionary
forces at the feet of the pyramids, overlooking the Nile.
England's overland route to India had thus been interrupted,
and 15 years of Franco-British wars of survival ensued.

The paramount role of Egypt in relation to Western com-
merce had become once again apparent when the plan by the
Frenchman Ferdinand de Lesseps to cut a waterway canal between the Mediterranean, through the Sinai Peninsula, to the
Red Sea, had evoked such a violent opposition in the Liberal
British cabinet under the leadership of Lord Palmerston.
The prospects of seeing France entrenched on a waterway
trade route across Egypt, which had all the obvious earmarks
of an open challenge to England's Atlantic route around the
Cape of Good Hope as well as to her overland trade route
across the Middle East, posed a severe threat to British
commerce. It was only after the Suez Canal, in spite of all
of England's efforts to the contrary, had become a reality,
that Conservative Prime Minister Benjamin Disraeli bought a
share of 4 percent interest in the stock of the Suez Canal
Company. In the following decade, the peace-loving, gospel-
reading Liberal Prime Minister William Ewart Gladstone ordered the bombardment of the city of Alexandria, which attack
in turn led to the beginning of 40 years of British occupation of the national territory of Egypt. Physical control
of the country, in spite of the great expense to the British
economy, had become an imperative foreign economic policy
for Great Britain in order to secure the international waterway passage through the Suez Canal.

The strategic role of the canal had emerged even more
sharply in the 20th century, at the time of Benito Mussolini'
invasion of Ethiopia and during the two world wars. However,
after World War II, the perpetual rivalries of Western powers
in Egypt "ceased for the first time in a hundred and fifty
years to be a relevant factor" in the consideration of their
foreign policies. A new balance of power heaved the postwar
world upon a seesaw of two polarized blocs of hostile nations
and Middle East policies were rapidly being remodeled in the
light of two major principles: the hostility of the USSR; an
the independent status, combined with the political, economic
and military weakness, of most of the states in the Middle
East.[14]

To Prime Minister Eden, the nationalization of the Suez
Canal thus loomed large upon the backdrop of British history.
It implied to him the extension of Soviet influence into the
growing political vacuum of the Middle East. In September
1956 Eden wrote in a letter to Eisenhower:

Similarly the seizure of the Suez Canal is,
we are convinced, the opening gambit in a
planned campaign designed by Nasser to expel
all Western influence and interests from Arab
countries.  He believes that if he can get
away with this, and if he can successfully
defy eighteen nations, his prestige in Arabia
will be so great that he will be able to mount
revolutions of young officers in Saudi Arabia,
Jordan, Syria and Iraq. . . .  These new Governments will in effect be Egyptian satellites
if not Russian ones.  They will have to place
their united oil resources under the control
of a united Arabia led by Egypt and under Russian influence.  When that moment comes Nasser
can deny oil to Western Europe and we shall be
at his mercy.[15]

Eden's friendly attitude toward U.S. decision-makers
was not shared by the French ministers.  Christian Pineau
held the United States directly responsible for the situation in Egypt.  "Nasser had described his action as retalia-
tion for the refusal to finance the Aswan Dam.  The United
States had a responsibility for this decision and should
not . . . disinterest itself from the consequences."[16]

But in Washington references to Middle East oil evoked
unsavory odors of antiquated colonial policies.  Liberals
and isolationists alike were standing by ready to accuse
the government of economic imperialism.  Most importantly,
the U.S. government itself was self-conscious of its position in relation to the new nations of Asia and Africa.  The
ink had scarcely dried on the South-Eastern Asia Treaty when
the emerging nations convened their own international conference at Bandung.  The Eisenhower Administration regarded
the Bandung Conference as a blow to SEATO, and great energy
was expended to mend fences.  Not only regional alliances
seemed to be threatened but the emergence of a unified Third
World bloc seemed to throw off balance all future votes in
the United Nations General Assembly.  Consequently, the political alignment of these nations became an issue of utmost
importance to both the United States and the Soviet Union.
On the U.S. scale of priorities, the Soviet specter loomed
most ominous over the Third World.  Any emerging nation,
unaligned with the "free world," posed a potential increment
of enemy power.  To prevent that from happening, the self-
styled leaders of the Third World were energetically courted.

Among these leaders, Nasser of Egypt, who was as vigorously
wooed by the Soviet Union as by the United States, was assumed to hold a prominent status. Any policy that might
alienate him and, under his influence, also other nations,
was condemned by the Department of State.

At this point, however, according to Eden, "M. Pineau
frankly regarded further talk with the United States Government . . . as a waste of time."[17] Thus, a military expedition evolved that shook the entire foundations of the North
Atlantic Treaty Organization.

To the Kremlin leadership, viewing the steadily increasing tensions between England and Egypt over the past 35
years, the nationalization of the canal by Nasser could have
seemed a welcome triumph, perhaps as important as the interruption of British trade routes, 150 years earlier, by Napoleon. NATO's forces depended for their mechanization and
mobilization on Middle East oil, and the Warsaw Treaty Organization could not but welcome this political coup. Even
if Egypt did not deny the use of the canal to the Western
nations, the contradiction in interests created by the nationalization threatened to undermine the cohesion of the
whole Western alliance.

The relationship between the United States and the
European members of the alliance had been described in pre-
ICBM literature as a relationship between a "shield" and a
"sword." This concept assigned to the European conventional
forces the role of "tripwires" on the way of the advancing
Soviet armies until the United States activated its nuclear
capability and thus swiftly put an end to the war while saving, in the process, the European cities and major industrial
centers from destruction by the enemy. Just what the relationship between the various members of the alliance should
be in a variety of other contingencies was not fully formalized by the allies. It was reasonable to expect, however,
that even if the United States did not bring forth its nuclear arms to bear on the situation, it, at least, would not
intervene in favor of the designated enemy.

But, while the U.S. government was still weighing its
sets of priorities in view of its global policy toward the
new and emerging nations of Asia and Africa, on the one hand,
and its position in relation to commitments undertaken in
respect to the other members of the NATO Alliance, on the
other, Nikolai Bulganin announced Moscow's determination "to
crush the aggressors by the use of force."[18] Thus, by
spreading the Soviet umbrella over the Suez zone, he stole
the U.S. thunder and drastically limited the range of options open to the U.S. government. Basically, Eisenhower

was now left with only two alternatives: He could counter
by spreading the nuclear umbrella over the Suez expedition
and thus calling Bulganin's bluff. Or, lacking evidence
that Moscow would indeed back down from its threatening position, he could avail himself of the Kremlin's well-known
propaganda line--the pious call to peace. President Eisenhower chose the second alternative.

His decision, however, was made too late to prevent the
petroleum shortages that the Western European governments
had dreaded so much since the nationalization of the canal.
Before the termination of hostilities put an end to the
grand fiasco of the Suez campaign, ships, already sunk in
the canal, blocked it to all sea-borne transportation while
the sabotaging, on Syrian territory, of pipelines belonging
to the Iraq Petroleum Company caused the interruption of
the overland flow of oil to the Eastern Mediterranean.
Europe was thus left with no direct access to Middle East
petroleum, and even the roundabout route via the Cape of
Good Hope could not be fully utilized in view of the great
shortages in the tonnage-carrying capacity of the tanker
fleets of the world.

## SHORTAGES OF PETROLEUM IN EUROPE

The seizure of the Suez Canal and its subsequent closure
by military action, together with the sabotaging of Iraq Petroleum Company pipelines, created a condition without parallel in the petroleum industry.[19] The Kirkuk-Tripoli pipeline
was only "the biggest single piece of Middle East production
that went out."[20] There were other interruptions, and their
effects were felt throughout the world.

The area most seriously affected by the disruption of
petroleum transportation was Western Europe. Postwar European demand for petroleum products continued to increase
steadily with the growing mechanization of the European
economies at about double the rate of increase in the United
States. Approximately two-thirds of the petroleum requirements of Western Europe were normally dependent upon the
Suez Canal and Middle East pipelines for transportation.[21]
Petroleum from the Middle East and the Far East to Eastern
Hemisphere ports west of the Suez, via the canal and the
pipelines, prior to the Middle East petroleum emergency,
amounted to approximately 2,165,000 barrels per day. About
two-thirds of this, 1,478,000 barrels per day (of which
50,000 barrels per day came from the Far East), moved through
the canal, and the balance, 687,000 barrels per day, moved

through the pipelines (Table 13). This was approximately 67 percent of the total requirements of 3,250,000 barrels per day for those areas; the balance of which, some 1,085,000 barrels per day or approximately 33 percent, was supplied from areas other than the Middle East and the Far East, such as the U.S. Gulf, the Caribbean petroleum area, Eastern bloc nations, and indigenous sources.[22]

TABLE 13

Oil Movement to West of Suez
(Eastern Hemisphere only)

| Route | | Amount (barrels a day) |
|---|---|---|
| Through Suez Canal | | |
| from Middle East | 1,428,000 | |
| from the Far East | 50,000 | 1,478,000 |
| Through pipelines | | 687,000 |
| | | 2,165,000 |
| From Western Hemisphere | | |
| and Soviet Bloc | | 1,085,000 |
| Total | | 3,250,000 |

Source: Based on information in U.S. Department of the Interior, Office of Oil and Gas, Report to the Secretary of the Interior, from the Director of the Voluntary Agreement, as Amended May 8, 1956, Relating to Foreign Petroleum Supply (Washington, D.C., June 30, 1957, mimeographed).

In summary, then, what was involved in practice in the petroleum emergency was how to keep a constant flow of 3,250,000 barrels per day going to Europe to meet its petroleum needs.

It was at this point in the Suez Canal crisis that the United States assumed full responsibilities for Europe's survival by structuring one of the most extraordinary worldwide organizations in all contemporary history. This effort, popularly known as the "Emergency Oil Lift to Europe," was perhaps second only to the Marshall Plan in the grandiose scale of its design and the magnitude of its operation.

The importance of this area for the national defense of the United States was stated by Assistant Secretary of the

Interior Felix E. Wormser, who was also the administrator
of the Voluntary Agreement Relating to Foreign Petroleum
Supply, as follows:

> . . . we have invested close to $50,000,000,000
> in the economic and industrial rehabilitation
> of our friends in Western Europe.  The invest-
> ment will be lost if our friends in Western
> Europe can not be supplied with petroleum prod-
> ucts; for this rehabilitation is based upon
> an economy in which the postwar increases in
> energy have been largely met by petroleum . . .
>   Further, we have given Western Europe
> large quantities of tanks, combat vehicles,
> motor transports, and aircraft.  All the im-
> plements of defense from the submarines under
> the water to the airplanes in the sky are
> fueled by petroleum, indispensable to every
> military operation.  Petroleum products not
> only move men and machines for the military,
> but the transportation systems that carry
> workers to the factories, move food from the
> farms to the cities, and transport manufac-
> tured goods from the factory to the con-
> sumers.[23]

To meet this demand for oil by the nations of Western
Europe, the United States organized a worldwide recruitment
of crude petroleum, petroleum products, and petroleum-
carrying tanker fleets, all of which participated in the gigantic effort to rescue Western Europe, for the second time
in less than five years, from an economic collapse due to
petroleum shortages.  The scarcity of petroleum was due,
however, not to any industrial shortages.  Western Europe
was short of petroleum not because the world lacked crude-
oil-producing-and-refining capacity but largely because not
enough transportation facilities existed to make up for the
loss of the Suez Canal waterway and the Iraq Petroleum Company's pipelines.[24]

## SOURCES OF ASSISTANCE

It was the opinion of the assistant secretary of the
Interior that "it was futile to mechanize Western Europe if
petroleum could not be made available to it."[25]  This view
was generally endorsed by all branches of the government.

The Department of the Interior estimated that 750,000 additional barrels a day of petroleum would have to be made available from all Western Hemisphere sources in order to meet the requirements of the Western European nations. "This quantity would be needed, in part, for direct shipment to the west of Suez areas affected by the closing of the Canal and, in part, to replace Middle East oil moving to North America which was assumed to be diverted to Europe."[26]

However, a subsequent analysis, made by Standard Oil of New Jersey, of the world supply/demand flow, including the available tanker fleet, indicated that under optimum conditions, Western Hemisphere sources of supply had to come up with an additional 105 million barrels daily if Europe's full requirements were to be met. The report weighed the practical aspects of the problem and emerged with the conclusions that (1) it would be reasonable to expect Europe, in view of the prevailing emergency conditions, to reduce its daily demand of oil by 15 percent and that (2) under the best conditions--that is, if tanker transportation with sufficient tonnage carrying capacity were available--"meeting Europe's restricted requirements was a real possibility."[27] This estimation was based on the availability of U.S. as well as worldwide petroleum resources. Thus, three major sources of supplies were counted on during the emergency: U.S. domestic production, oil from the Caribbean area, and Middle East oil diverted from its normal course in the Western Hemisphere.

Since the administration of Calvin Coolidge, under U.S. petroleum conservation policy, there had always been a policy, though not always fully implemented, of maintaining a constant supply of crude oil available for a period of 10 to 12 years. In spite of petroleum panics that visited the country periodically, every year there was a little more oil found in the United States than there was oil consumed.[28] In order to meet the prevailing emergency conditions in Western Europe, domestic petroleum production in the United States was increased by approximately 300,000 barrels per day as of November 1, 1956.[29] During this period, approximately 235,000 barrels of crude oil were shipped from the United States to Western Europe each day. Of this amount, 150,000 to 200,000 barrels came out of the new production.[30] The overall average shipped from the United States to Western Europe was about 475,000 barrels per day including crude petroleum and petroleum products.[31] This constituted a remarkable increase not only of shipment of crude petroleum but also of export of petroleum products to Western Europe.

Before the Middle East emergency the United States had
shipped to Western Europe only 50,000 barrels of petroleum
supply each day.  Of the average 475,000 barrels shipped
from the United States to Western Europe during the emergency, 235,000 barrels were crude oil and 230,000 barrels
were petroleum products.

Many parts of the domestic industry and many government
agencies and military services participated in the emergency
effort.  The Military Sea Transportation Service, for example,
reactivated its reserve and mothball tanker fleets, while
the United States Coast Guard waived the load limitations on
U.S. flag tankers engaged in coastal trade.[32]  U.S. assistance to Western Europe in its time of most acute oil shortages was subject to four legal limitations:  those set by
the existing conservation policy of the United States; those
set by the Defense Production Act; those set by the Voluntary Agreement; and those set by the ad hoc Middle East
Emergency Committee (MEEC).

Each one of these limitations, in turn, is dealt with
in the following pages.


THE CONSERVATION POLICY OF THE UNITED STATES

World War I had pointed out the immense importance of
raw materials to national defense and the crises that could
befall the entire national economy in their absence.  The
significance of oil to the national interest had been height-
ened by its contribution to the Allies' victory over the
Central Powers.  The postwar concern with anticipated shortages of oil reserves was augmented by an increasing dependence of the United States on an inflow of foreign oil, for,
beginning in the 1920s, the United States ceased to be a net
exporter of crude petroleum and became a net importer.  The
anxiety over the adequacy of future reserves spurred a steady
buildup of excess supply of petroleum, a policy that soon
became a controversial issue as to the utility and economy
of the method employed.

In view of the great strategic value of petroleum and
the rapid development of aviation during that decade and
taking into consideration the fact that petroleum was the
key to national prosperity, President Calvin Coolidge appointed, on December 19, 1926, a Federal Oil Conservation
Board whose task was to study and prepare recommendations as
to the best and least wasteful methods to conserve the oil
resources of the nation.  The board prepared several reports,
which it submitted to the president, and studied a great number of memoranda that it collected from various sources.

The real contribution of the Federal Oil Conservation Board to policy-making was its preparation of a program of statistical information and forecasts as to national "demand for refined petroleum products and through it the amount of crude necessary to yield the products so demanded."[33] By administrative order, dated March 10, 1930, issued by Secretary of the Interior Ray Lyman Wilbur, this procedure became the blueprint according to which conservation policies have been formulated to this day. At present, the Bureau of Mines is responsible for the collection of all pertinent data.

Two schools emerged in Washington that offered opposing solutions to the problem of how best to conserve the excess capacity of oil production in the United States. Both agreed that the well-known strategic property of oil made its immediate availability imperative. But each recommended a different estimate of the utility of oil stockpiling for time of emergency. According to one view, setting aside areas assumed to contain such reserves was "a wholly inadequate measure of preparing for war emergencies. . . . If the oil was to be available when needed for war emergencies, the work that would assure its availability must be undertaken long before the emergency arose."[34] This view required that expenditures go to actual stockpiling of oil reserves on the land surface.

The opposing view was expounded by J. E. Warren, who was the deputy administrator of the Petroleum Administration for Defense:*

> You can't stockpile petroleum for an emergency as we are stockpiling lead and zinc and aluminum and some 70 other essential materials. It would be fabulously expensive and excessively wasteful to pump oil out of the ground into storage tanks and let the oil sit there against the day of possible need. There is only one truly satisfactory stockpile of petroleum, and that is an underground stockpile of available possible production beyond the actual production on any given day.[35]

---

*His statement, however, was not borne out by the facts, as demonstrated in 1940, when the United States did have reserves of this sort only because a number of large oil fields had been recently discovered in a relatively short time.

This view became the basis of U.S. petroleum conservation policy. The Petroleum Conservation Commission was assigned the task of ascertaining that at any given time
underground oil reserves would be available in the United
States, explored and ready to be drilled, prepared for immediate production, in quantities sufficient to meet the
normal future civilian and military demands of the nation,
as forecast, for a period extending over the succeeding 10
to 12 years. It was this source of discovered existing reserves upon which the United States drew to supplement much
of Western Europe's short supplies during the Suez Canal
petroleum crisis. In the process, these reserves were
depleted far below the minimum requirement stipulated by
law, and thus was jeopardized the entire preparedness posture of the United States in case of another emergency.
Moreover, not only were civilian reserves drained because
of the energy crisis in Europe but military reserves in the
United States were also drawn upon, with very serious results for the mechanization potential of the armed forces.


## THE DEFENSE PRODUCTION ACT

As during World War I, so also in World War II, the
availability of petroleum was a decisive factor in the Allies' victory. Its significance to the strength of national
economies was soon demonstrated in its contribution to the
reconstruction of Western Europe under the Marshall Plan.
But at the same time, when European nations were devoting
all their energies to economic recovery, new conditions of
international tension were created under the impact of the
sharply polarized balance of power and the resultant Cold
War. The precarious equilibrium deepened the political dimensions of the international dilemma of oil. When, in
1950, war broke out in the Far East and strategic materials
once again became the paramount security assets of the nation, Congress was called upon to enact legislation that
would bring much of this vital industrial capacity under
limited but effective government control and give it material support.
     In enacting the Defense Production Act of 1950, Con-
gress recognized the need for exploration and development of
the mineral resources of the country and that the actual
work should be done by private enterprise. The act also
recognized that where private capital could not assume the
long-term burden of exploration, financial assistance by the
government was necessary in a program for the encouragement

of discovery and development of mineral reserves, particularly those in short supply and critically needed for the national defense program and basic civilian requirements. The act provided much of the basic authority to bring about expansion of productive capacity, provide controls over the use of scarce materials, and initiate other measures essential to enhance military strength. The defense materials system provided an important readiness measure. According to testimony before the House Committee on Banking and Currency,

> It was a device which could be expanded rapidly in time of emergency to handle the abrupt increase in military and defense-supporting orders which would occur. It provided the framework of the more detailed measures, which would become essential to control the use of scarce materials in the civilian markets.[36]

Defending the legitimacy of government intervention in free market practice, President Truman, in his message to Congress of February 11, 1951, asserted that this act was "essential to the defense mobilization effort of the Nation." It was, he said,

> a program to create invincible defense strength in the free world . . . an authority to channel materials for defense, to help expand essential production, and to help small business make its vital contribution to the mobilization effort . . . an authority to stabilize prices, wages, credit, and rents. . . . Only strong controls can give businessmen and consumers assurance that prices will not be allowed to get out of hand, and that there is no need for panic buying.[37]

Under this act, activities since the Korean War have resulted in the addition of many new production facilities and supplies of materials, which contributed greatly to the nation's ability to mobilize quickly. As the broad expansion program progressed, the Office of Defense Mobilization was able to give greater consideration to "the more specific bottlenecks to rapid conversion to a war footing."

Of special interest to the petroleum industry was Section 708 of the Defense Production Act. This section authorized the granting of "an exemption from the antitrust laws for private parties voluntarily engaging in combined action found to be in the public interest as contributing to the national defense." The second substantive power in Title VII was found in Section 710 of the act, which was added in 1955. This section authorized the executive reserve program. According to this provision, "each agency having mobilization responsibilities is authorized to set up and train a reserve of persons from private life and from Government to fill executive positions in the Federal Government in time of mobilization."[38] The Department of Commerce concurred. In its estimation, this was a "tremendously important program because it assures us of the ready availability of trained manpower without the necessity of keeping a number of individuals on the Federal payroll for this purpose."[39]

The Defense Production Act thus established integration committees to "meet to exchange information, techniques and processes." Under the existing law, the voluntary agreements, under which these committees operated, were the best way to deal with various types of emergency situations. Unless the government wanted to get much more deeply involved in industrial operations, these provisions in the Defense Production Act provided the best method for dealing with a situation such as arose in regard to Middle East petroleum.[40] It provided both a framework and a basis, rooted in defense requirements of the country, for the actual support of the domestic petroleum industry. Nevertheless, neither during the Iranian petroleum crisis nor at the time of the Suez Canal crisis was the United States prepared to meet the emergency requirements with sufficient quantities of available petroleum reserves. The Voluntary Agreement Relating to Foreign Petroleum Supply, through its Plan of Action, established the ad hoc Middle East Emergency Committee, and supplementary sources were, indeed, quickly located in other parts of the world. But the short-sightedness of the tanker-building policy of the United States made the transportation of such supplies to places of most acute shortages a major technical handicap and thereby created the real Suez Canal petroleum crisis.


THE VOLUNTARY AGREEMENT RELATING TO
FOREIGN PETROLEUM SUPPLY


The Defense Production Act is reenacted periodically by Congress, at which time it is subject to investigation by congressional committees of both houses, which may recommend additions, deletions, revisions, and amendments. In approaching the problem posed by the petroleum crisis of 1956-57, the U.S. government had to rely on the existing Voluntary Agreement Relating to Foreign Oil Supply as stipulated under the provisions of the Defense Production Act as amended in 1955. The provisions of Section 708 (Title VII) of the act as amended provided that new voluntary agreements or programs could not be entered into after the enactment of the amendments of 1955, unless they were made by defense contractors with respect to other military production. Accordingly, shipments of oil and oil products to the NATO contingents in Europe could, conceivably, fall under the category provided for by the act as amended. The largest part of the emergency oil lift to Europe, however, was shipped to local distributors for civilian requirements The letter of the law, then, forbade this kind of operation and opened a door to questioning the legality of the 1956 agreement and the committees established by it.

When questioned on this point of the law by Senator Joseph C. O'Mahoney, Charles H. Kendall, general counsel to the Office of Defense Mobilization, stated that the Volunta Agreement of 1956 "was not an agreement at all, properly. is a plan of action under the original agreement"[41] entered into prior to the 1955 amendment. In other words, the Department of the Interior and the Office of Defense Mobiliza tion now considered the Voluntary Agreement Relating to For eign Petroleum Supply of 1953 as the basic and binding instrument, while any subsequent agreements entered into by the same parties were merely revisions of the original contract.* This argument was further enhanced by the evidence that clearance sought at the office of the attorney general in regard to antitrust provisions had been approved on July 20, 1953--long before the 1955 amendments--and was equally applied to subsequent agreements. Consequently, th Defense Production Act as amended in 1955 was not being violated at all, as the emergency oil lift program was really based upon an agreement that preceded the enactment of the new amendments and therefore was valid in the eyes of the law.

---

*The voluntary agreement of May 1, 1953, superseded th temporary agreement entered into in 1950 as a result of the Iranian oil crisis. It was amended once since then, on April 15, 1954, at which time it became operative for the p riod commencing on July 31, 1955 and ending on March 31, 19

Thus, on May 8, 1956, when the U.S. government was already anticipating difficulties in the Middle East as a result of the strained relations with Egypt, an amended Voluntary Agreement Relating to Foreign Petroleum Supply came into being. According to rules established by the precedents of former agreements, a standing Foreign Petroleum Supply Committee, composed of government employees and industry representatives, was set up in order to supervise the operative aspects of the agreement. The creation, on August 10, 1956, of an ad hoc Middle East Emergency Committee, whose task was to execute the FPSC's emergency Plan of Action relating to the Suez Canal petroleum crisis, raised the question of the validity of the MEEC under the amended Section 708 of the Defense Production Act, in view of the fact that this new agreement was entered into by nonmilitary contractors for, largely, nonmilitary purposes. But the Department of the Interior denied this charge. It claimed that the new agreement in regard to the ad hoc committee was merely an operative extension of the Voluntary Agreement Relating to Foreign Petroleum Supply as legally recognized by the Office of the Attorney General and other branches of the government in 1953. The Plan of Action became henceforth the operative name designated by the Department of the Interior to the MEEC, in order to sidestep any further legalistic allegations that this committee had been set up in violation of the amended Section 708 of the Defense Production Act of 1955.

The legality of either committee or any of the agreements was never seriously challenged by either the Department of Justice or Congress in respect to the validity of the committees or agreements under the Defense Production Act as amended in 1955. However, questions were raised, as we shall see, regarding antitrust violations.


## THE WORKINGS OF MEEC

On May 8, 1956, the amended Voluntary Agreement Relating to Foreign Petroleum Supply and its operative arm, the Foreign Petroleum Supply Committee, came into being. After the Suez Canal was nationalized on July 26, Secretary of the Interior Fred A. Seaton and Director of the Office of Defense Mobilization Dr. Arthur S. Flemming reached an agreement on the necessity for preparation of a plan in the event of "possible emergencies."[42] A proposal was then made to organize, under the Voluntary Agreement Relating to Foreign Petroleum Supply, an ad hoc Middle East Emergency Committee

composed of government officials and representatives of
U.S. oil companies engaged in foreign petroleum operations.
The proposal to bring such a committee into existence was
recommended by the Department of the Interior and was concurred in by the secretary of State, the attorney general,
and the Federal Trade Commission; it was approved by the
director of Defense Mobilization.[43] Hugh Stewart, director
of the Office of Oil and Gas and chairman of the Foreign
Petroleum Supply Committee under the Voluntary Agreement,
was requested to prepare such a plan. On August 10, the
Middle East Plan of Action was approved. No congressional
action was sought or taken then or later on the Plan.[44]
Likewise, the international petroleum industry was ignorant
of it. Stewart P. Coleman, vice president of Standard Oil
Company of New Jersey, received the information about the
Plan's existence from the government on July 31,[45] after it
had already been created, at which time he was invited to
be chairman of the Plan of Action in charge of its operations. The Plan had been conceived by members of the Office
of Oil and Gas at the Department of the Interior, without
any consultation with or pressures by Congress or the industry. It was Seation's intention to place the entire emergency operation completely under government direction.
Within three months from the date of approval of the Plan
of Action, on October 31, the Suez Canal was blocked; on
November 3, the pipelines belonging to the Iraq Petroleum
Company were sabotaged on Syrian territory. At the end of
that month, on November 30, the Plan of Action was put into
motion.

    MEEC was set up, apart from the standing Foreign Petro-
leum Supply Committee, in order to execute the operative end
of the Plan of Action. The principles and procedures of
MEEC were not new. They had formed the basis for coordinated oil industry effort during World War II. Their outline and general objectives were incorporated in the two
reports submitted by the National Petroleum Council to Secretary of the Interior Krug on January 13, 1949. They were
again employed at the time of the Korean War as a result of
the Iranian petroleum crisis. In the opinion of the Department of the Interior, "to have used an unproven method would
have been to gamble with the economic stability of the Free
World, particularly of the European nations in the NATO alliance."[46]

    All provisions specified in the Voluntary Agreement
Relating to Foreign Petroleum Supply applied equally to the
standing Foreign Petroleum Supply Committee and MEEC. Under
these provisions, all activities of the committee and its
subcommittees were under the supervision and direction of

two government officials, the administrator and the director, both of whom were designated for their positions by the secretary of the Interior. Thus, the operation of MEEC and its subcommittees was under immediate supervision, direction, and control of the government represented by the administrator and the director and their staff of assistants recruited from the Office of Oil and Gas at the Department of the Interior. The elements of government control consisted of the following provisions: (1) MEEC and its subcommittees could act only on the request of the administrator or the director; (2) meetings of MEEC and its subcommittees could be held only when called by the administrator or director; (3) the agenda of all meetings were prescribed in advance by the administrator or director, and copies of the agenda and notices of the meetings were sent to the Office of the Attorney General; (4) all meetings had to be attended by at least two full-time government employees; (5) the meetings were open to representatives of any government agency, representatives of the Department of Justice, and representatives of the Department of the Interior; (6) no other persons, except for those specified, were permitted to attend any meeting except as designated by the administrator or the director; (7) all meetings had to be held on government property; (8) all actions taken were first to be cleared by the Plan of Action Review Committee; (9) the member companies were required to report periodically to the director concerning their activities in connection with certain foreign governments and industrial committees in Europe that were dealing with the stoppage from their end. All these reports were immediately forwarded by the administrator to the Department of Justice for review.[47]

The Plan of Action also stipulated several important limitations on the scope of activities and powers of MEEC: the committee had no power of compulsion over anyone; it had no power or authority to allocate petroleum supplies; it had no power or authority to enter into or operate pooling arrangements of any kind, including tanker pools; it had no power or authority over domestic petroleum operations; it had no power or authority to buy or sell oil or to make deals for the purchase or sale of oil; it had no power or authority over the prices of crude oil or its products; it had no power or authority with respect to tanker rates; and it had no power or authority with respect to the commercial or financial aspects of actions by its members or other persons.

According to the Interior Department report, "The Plan of Action was designed to prevent, eliminate or alleviate shortages of petroleum supplies which threatened to adversely affect the defense mobilization interests of programs of the United States."[48]  The objective of the Plan of Action was to devise measures "to offset or minimize petroleum shortages and dislocations in friendly foreign areas brought about by a substantial Middle East petroleum transport stoppage."[49] The original objective of the Plan of Action was to supply Europe 100 percent of its normal demand, "not necessarily from the United States, but . . . using all available sources of supply, bearing in mind the limitation that we have of tanker transportation."[50]  The Office of Oil and Gas estimated that the United States was in a good position to supply Europe with 1.1 million barrels per day, in order to support Europe's "defense against Communist aggression" by keeping Europe economically "strong and in a position to continue to serve as effective allies."[51]  However, very early on, the committee recognized that this objective was unattainable. The international petroleum industry's specified goal was to satisfy 85 percent of Europe's normal daily demand of petroleum and petroleum products, and the industry expected the member nations of the OEEC and Spain to modify their demands accordingly for the duration of the emergency.  Coleman, the MEEC chairman, promised to meet better than 80 percent of Europe's normal demands, and even more of the normal fuel oil demand.[52]  However, one provision of the Plan of Action stipulated that there should be no deficiency in U.S. supplies while pooling oil resources and tanker facilities. To this end, MEEC was given the responsibility of supervising all these diversified activities while seeing to it that there was no reduction in available supplies to this country.

MEEC was composed of government employees and representatives of those U.S. oil companies that, in the aggregate, had extensive operations in the foreign areas immediately affected as well as throughout the rest of the world.  Fifteen international petroleum companies initially participated in MEEC.  These 15 companies were properly designated as the largest in the world.  Their total assets amounted to $20,162,034,000 in 1955.  Initially, the government chose to limit the membership of the committee to those companies that had the largest overseas operations.  Later, a few companies having smaller investments abroad and companies having only domestic operations were admitted.  The reason, at first, for inviting only companies with foreign operations was, primarily, the possession by them of sea-borne transportation facilities, of which the U.S. government had scarcely enough and domestic producers had none.

The responsibilities of MEEC were distributed among seven subcommittees:  the Supply and Distribution Subcommittee; the Tanker Transportation Subcommittee; the Production

Subcommittee; the Refining Subcommittee; the Pipeline Transportation Subcommittee; the Statistical Subcommittee; and
the Information Subcommittee.

On January 17, 1957, the administrator established the
Plan of Action Review Committee, an intergovernmental group
whose task was "to review all actions taken by the partici-
pating companies in implementation of the approved sched-
ules."[53] This three-men committee was composed of one repre-
sentative each from the Office of Defense Mobilization, the
Department of Justice, and the Department of the Interior.
The director of the Voluntary Agreement Relating to Foreign
Petroleum Supply, who was also the representative for the
Department of the Interior, was chairman of the committee.

The history of MEEC falls into three distinct phases:

First Phase:  from August 24, 1956, to December 3, 1956.
During this period, MEEC and its subcommittees were "engaged
exclusively in the preparation of information and data concerning foreign petroleum operations."[54]  No schedules were
actually prepared until after the Suez closure, but planning
data were assembled and analyzed.[55]

Second Phase:  from December 3, 1956, to April 18, 1957.
This was the critical period of the Middle East petroleum
crisis.  Practically all the actions devised to offset the
supply shortages in the Eastern Hemisphere west of Suez were
taken during this phase.[56]

Third Phase:  from April 18, 1957, to June 30, 1957.
At this time the activities of MEEC and its subcommittees
became confined to the preparation of data on the status of
petroleum supply in the affected areas and to complete the
historical record of the petroleum movements during the
period of crisis.[57]

The activities of MEEC and all its subcommittees under
the Plan of Action terminated as of July 31, 1957.

The period of actual emergency resulting from the Suez
Canal crisis lasted, generally, from November 1, 1956,
through March 31, 1957.  During this period an average of
3,007,000 barrels per day, or better than 90 percent of
Europe's daily requirements, were made available to the
affected areas.[58]  "This rate of supply continued through
April and May, 1957, and in June, 1957, total supplies to
the affected areas, as well as movements to these areas
through the Suez Canal, returned, for all practical purposes, to normal proportions."[59]

Anticipating this large-scale redistribution of the
world's petroleum production, MEEC prepared, during its
first phase of operations, a series of five schedules that
diverted the tanker fleets from their normal course to ports
of the affected areas as follows:

| Schedule | | Administrator's Action 12/7/1956 |
|---|---|---|
| No. 1 | five major diversionary movements | approved |
| No. 2 | three major diversionary movements | approved |
| No. 3 | rearrangement of petroleum supplies movements east of Suez | approved |
| No. 4 | rearrangement of pipeline and other petroleum transportation in U.S. | approved |
| No. 5 | rearrangement of petroleum transportation from U.S. and Caribbean to South America | not approved (2/11/1957) |

The operation of all these schedules was subject to two conditions: that they were reasonably calculated to reduce petroleum shortages in friendly foreign nations, which arose as a result of the Middle East emergency, and that they would not cause a deficiency of petroleum supplies in the United States.[60] A major outgrowth of the operation was the "increase in the effective carrying capacity of the existing tanker fleet made possible by radical redeployments of tanker movements designed to eliminate cross hauls and unusually long hauls."[61]

Optimistic estimations of MEEC's work were not shared by everyone. In mid-January it became apparent that shipments were inadequate. Members of Congress received numerous complaints about the increased cost of petroleum products. In February the Senate Committee of the Judiciary and the Senate Committee on Interior and Insular Affairs jointly probed through their subcommittees into the causes for the price hike of petroleum and petroleum products and into possible violations by the oil companies of the antitrust laws of the United States in respect to monopoly and price-fixing practices. Under the cochairmanship of Senators Estes Kefauver and Joseph C. O'Mahoney, the investigating subcommittees addressed themselves to four principal categories of inquiry: (1) Why had the oil companies not sent the planned amount of oil to Europe? (2) Why had the oil companies nevertheless raised the price of oil in Europe? (3) Why had the oil companies raised domestic prices also? (4) In spite of this price rise, why had very little more oil been produced, despite huge oil inventories?[62]

142

But to O'Mahoney the principal objective of the investigation was to qualify the privileged status granted the petroleum companies under provisions of the voluntary agreement.

> The mere recitation of the commingling of foreign political and American private authority engaged in the exploitation of the petroleum deposits of the Middle East, raises the central question at issue in these hearings, namely, whether such a group of oil companies is qualified to be chosen by the Government of the United States, under the Defense Production Act of 1950, as amended, to carry out a policy which this Government is undertaking for purposes of national defense. . . .
> When, therefore, protests are received by this committee, when complaints are made of price increases, and when, as is the case here, the members of the Middle East Emergency Committee include corporations which have already been sued by the Federal Government for allegedly participating in illegal foreign cartels, it becomes the obvious duty of this committee, as well as other committees of the Congress, to scrutinize carefully the formulation, the procedures, and the effect of the plan.[63]

Obviously, a situation favorable to the creation of monopoly had developed as a result of the organization plan of the MEEC. This view was expressed most strongly by O'Mahoney, who charged "that the fifteen giant corporations which make up the original members of the M.E.E.C. have a selfish interest when they undertake to manage for the Government the oil lift program,"[64] especially in view of the fact that they were granted some immunity from antitrust laws. O'Mahoney articulated, to a large degree, some real and imagined grievances of the domestic petroleum industry, whose "primary concern with respect to the M.E.E.C. was that the few individual companies comprising this committee would not be given any authority, with antitrust immunity, to take any actions with respect to or which would adversely affect the petroleum industry within the United States."[65] To correct this imbalance of influence of the few over the petroleum policies of the United States, the independent U.S. producers suggested that the National Petroleum Council

"could have presented to the Government a far more effective
plan than the one now in operation, and one which would not
have required the granting of antitrust immunity."[66]

Such a plan, however, while helping to safeguard the
interests of the domestic producers, could not possibly meet
the worldwide requirements of the government in recruiting
all available petroleum supplies and transportation facilities.  It would have been disastrously impractical on the
part of the government to entrust the entire oil-lift program to the domestic producers, since even the little that
they had been asked to do they did not do.  "Even though the
President had asked that prices be held down, they did nothing to implement this request."[67]  In fact, in the midst of
the emergency effort, prices of petroleum and all its products were raised.  The Office of the Attorney General received complaints that this price rise was a result of collusion, and a grand jury action against the domestic independent petroleum companies began in early February.

President Eisenhower had indeed asked both business in
general and labor to hold the price and wage line as much as
possible.  In doing so the president followed a well-
established pattern in U.S. price and wage history, as similar appeals had been made to industry as a whole and to
specific industries in particular in the early days of World
War II and the Korean War, prior to enactment of control
legislation.  However, no formal freeze was initiated by the
Eisenhower Administration.  It was the opinion of the administration that ordering companies to hold down prices would
constitute an interference with the free market.[68]

The domestic price hike was matched several times over
in the affected areas of Western Europe.  The tanker rates
nearly doubled because the owners had taken advantage of the
war clauses in their contracts.  When a shooting war had
started at Suez, they immediately canceled their contracts
with the producing companies and demanded from them higher
rates.  No freeze policy could be applicable in this instance, as the United States, not having any sizable merchant marine of its own, had to rely on fleets of foreign
flags.  If tanker rates were to be frozen, then, it could
not be done by the United States alone.  In order to be effective, freezing would have required the cooperation of all
governments involved in the emergency program.  No such cooperation could be accomplished within a short period, since
many tanker fleets were flying flags of convenience whose
governments were not directly affected by the Middle East
crisis.  Thus, due to the short-sighted policies of the U.S.
government in failing to replenish its merchant marine after

World War II, the administration became entirely dependent
for its transportation facilities on the acute business sense
of private contractors in their dealings with foreign governments.  Moreover, because of this default of the merchant
marine, U.S. assistance to the OEEC nations and Spain cost
these countries a large portion of their well-accumulated
balance-of-payments dollar reserves.  Thus, while rescuing
European industry from a standstill because of lack of sufficient petroleum supplies, the United States practically
pulled the rug from under the European economies by depleting
their coffers.

## THE AFTERMATH OF THE SUEZ CANAL
## PETROLEUM CRISIS

The lesson of the Suez crisis was swift and bitter.  As
will presently be seen, it discredited past policies of the
Eisenhower Administration and brought about a shift toward
new objectives.

The U.S. government had, initially, approached the
problems raised by the Middle East events under the impact
made on it by the recently articulated doctrine of noncommit-
ment at the Bandung Conference of "neutral" Afro-Asian nations and the rapidly increasing number of such voting member nations in the United Nations General Assembly.  Five
heads of state seemed to have focused attention upon themselves at that conference:  Nasser of Egypt, Jawaharlal Nehru
of India, Kwame Nkruma of Ghana, Sukarno of Indonesia, and
Josip Tito of Yugoslavia.  The aim of these Third World
leaders was to be a force for peace by remaining politically
uncommitted to either bloc and keeping themselves aloof from
the Cold War.

This apolitical approach to international relations con-
noted to the first Eisenhower Administration all sorts of
ominous possibilities.  Generally, it was agreed that an
apolitical posture signified, in effect, a political vacuum
in that area that could be skillfully turned into a fertile
ground for all sorts of communist manipulations.  It became
the first order of business for the U.S. government to bring
these nations into a closer alignment with the Western world.
Therefore, the Anglo-French attack upon Egypt in the region
of the Suez Canal was considered by the State Department to
be ill-timed and ill-advised.  The State Department opposed
not only a military attack upon the territory of Egypt but
also any actions that would cause the alienation of the
Third World, moved by Nasser's eloquence, from the Western

alliance. Incurring this kind of organized hostility was weighed against the utilitarian purposes of regaining the international legal status of the Suez Canal.

This view of the unaligned nations as a unified world-wide force shaped U.S. Middle East policy during the Suez Canal crisis. But very soon it became apparent that aside from a number of letters and draft resolutions at the United Nations General Assembly,[69] the uncommitted nations of Asia and Africa had demonstrated neither solidarity with their sister nation on the Red Sea nor any kind of uniform policy in regard to the incident. Instantly, the myth of the Third World was shattered.

After the Suez fiasco the State Department tended to view the Bandung Conference as a platform of pompous prom-ises, and Third World now became a hollow term. Moreover, the Middle East, because of its demonstrated political and military weakness, was viewed, in the aftermath of Suez, as an increasingly sensitive crossroads of Cold War strategies where Western prestige had already been dealt a blow by So-viet intimidation. The inability of Middle East nations to defend themselves against an external attack without the assistance extended by Western commitments was undisputed. However, such Western commitments as there existed at the time were being radically reduced. The French presence in the Middle East (not including North Africa) had ended dur-ing World War II, while British commitments in that region were rapidly narrowing down to a few scattered bases. At the same time, the two existing political instruments, the Tripartite Declaration of 1950 and the Baghdad Pact of 1955, did not long survive the Suez Canal crisis.

The Tripartite Declaration was signed by the govern-ments of France, the United Kingdom, and the United States in an effort to curb the arms race in the Middle East and was designed to apply equally to all Middle East nations to assure them of their independence and territorial integrity. The Anglo-French attack upon the territory of Egypt seemed to have nullified all meaningful content of that agreement. However, on May 20, 1958, Secretary of State John Foster Dulles was still upholding this treaty:

> We do regard it as applicable. We don't re-
> gard it as powerful, you might say, . . .
> because that Tripartite Declaration has never
> had specific Congressional approval. We have
> always considered that whether action under
> that, or another Declaration that President
> Eisenhower made, I think, in '56, dealing

with these problems, that the Constitutional
power of the President to act under those was
not as great as though they had received ex-
press Congressional approval.[70]

The Baghdad Pact of Mutual Cooperation was originally
signed between Iraq and Turkey on February 24, 1955.  On
April 4-5, 1955, Great Britain became a party to it through
a special agreement signed between it and Iraq.  Pakistan
signed on September 23, 1955, and Iran on November 3, 1955.
The pact recognized the great responsibilities concerning
the maintenance of peace and security in the Middle East
region but made no provisions for armed intervention to
deal with subversion.  On April 18, 1956, the United States
became a full member of the Economic Committee of the Baghdad
Pact.  However, in its statement in support of the Baghdad
Pact, the U.S. government revealed itself to be interested
in the political rather than the economic aspects of the
pact:

The United States reaffirms its support for
the collective efforts of these nations to
maintain their independence.  A threat to
the territorial integrity or political independence of the members would be viewed by
the United States with the utmost gravity.[71]

The U.S. willingness to undertake the military as well
as economic commitments implied in the pact received official
recognition when the U.S. government, hoping to breathe new
life into the old instrument, signed on March 5, 1959, a
series of bilateral executive agreements with Iran, Pakistan, and Turkey, all members of the pact.  These agreements,
undertaken in accordance with purposes enunciated by the
Eisenhower Doctrine, were "tantamount to substantive, if not
formal, membership in the Baghdad Pact."  In the same month,
Iraq withdrew from the pact, and the headquarters was transferred from Baghdad to Ankara and the name was changed to
Central Treaty Organization (CENTO).
In addition, there existed a bilateral agreement between Egypt and Great Britain.  Signed in 1954, the Anglo-
Egyptian treaty provided for the evacuation of British
troops from the Suez Canal Zone, subject to a provision for
their return to the canal base in the event that there existed a danger to the territorial integrity of Egypt.  But
on January 1, 1957, following the Anglo-French attack, Egypt
denounced this agreement, and on February 13, 1957, Great
Britain recognized its termination.

In the eyes of the U.S. government, neither the Tripartite Declaration nor the Baghdad Pact seemed to provide
sufficient guarantee to the security requirements of the
Middle East region.  The Eisenhower Doctrine[72] was enunciated in order to create that umbrella.  Its power stemmed
from the fact that, unlike the other two instruments, it received congressional approval and, therefore, was considered to be a stronger mandate, which gave the executive "a
greater authority than if it would have been a declaration
by the President himself."[73]  The Eisenhower Doctrine essentially reflected U.S. recognition that "as a result of the
Suez operation it had inherited, to a considerable extent,
the responsibility for the Western position in the Middle
East."[74]  As Prime Minister Harold Macmillan pointed out in
Parliament, Eisenhower's message to Congress, and the proposals it embodied, was designed to provide a legislative
basis for future action.[75]  Secretary of State Dulles repeated very much the same view when he said in a press conference that "the Middle East resolution has received congressional approval and, therefore, we consider that it is
a stronger mandate and it gives the President a greater
authority than if it would purely have been a declaration
by the President himself."[76]

At the same time, there was a growing realization in
many quarters that while Europe's cheapest sources of energy
supply were derived from the Middle East and North Africa,
"most of this oil, except for Iranian oil production, was
actually or potentially Arab oil, and therefore . . . this
oil supply was particularly subject to hostile political
control."[77]

This fear of outside blackmail due to dependence on
Arab oil was offered by Eisenhower as the basic reason for
new directions in future Middle East oil policy.  He stated
that the Middle East

> contains about two thirds of the presently
> known oil deposits of the world and it nor-
> mally supplies the petroleum needs of many
> nations of Europe, Asia and Africa.  The na-
> tions of Europe are particularly dependent
> upon this supply, and this dependency re-
> lates to transportation as well as to pro-
> duction.  This has been vividly demonstrated
> since the closing of the Suez Canal and some
> of the pipelines. Alternate ways of trans-
> portation and, indeed, alternate sources of
> power can, if necessary, be developed. But
> these cannot be considered as early prospects.

These things stress the immense importance of the Middle East.  If nations of that
area should lose their independence, if they
were dominated by alien forces hostile to
freedom, that would be both a tragedy for the
area and for many other free nations whose
economic life would be subject to near strangulation.  Western Europe would be endangered
just as though there had been no Marshall
Plan, no North Atlantic Organization.  The
free nations of Asia and Africa, too, would
be placed in serious jeopardy.  And the countries of the Middle East would lose the markets upon which their economies depend.  All
this would have the most adverse, if not
disastrous, effect upon our own nation's economic life and political prospects.[78]

The new shift in U.S. Middle East policy was initiated.
It was brought about not merely because no other Western
power could assume large-scale military commitments there
anymore but primarily because the Soviet presence in the region had already become a paramount concern of the NATO alliance.  Communist arms deals with the Arab nations prior to
the Suez Canal crisis and the promptness with which the Soviet Union replaced Egyptian losses of materiel immediately
after the 1956 war gave rise to a belief that, unless measures were taken to curb Soviet influence in those countries,
much of that area might soon be slipping into a hostile camp.
Soviet penetration into the Middle East region was conducted on several planes:  open military assistance, covert
political activities, and new official foreign trade policies.  Peaceful coexistence with economic competition became
the watchword of this unprecedented Soviet trade offensive
into the "free world."  The principal weapon in the Soviet
foreign trade arsenal was the abundance of petroleum resources that the USSR possessed.  Under the leadership of
Premier Nikita Khrushchev, Soviet foreign trade in petroleum
penetrated the far-flung corners of the earth, bringing Soviet influence to bear on near and distant economies.
Khrushchev had turned petroleum into a political weapon,
and Eisenhower countered by enacting his own doctrine, which
was designed to communicate a veto message to any further
Soviet incursions into the Middle East.  The new legislation
was "primarily designed to deal with the possibility of Communist aggression, direct, and indirect."[79]  At the same
time, efforts were being made not only to restructure alternate ways and means of petroleum transportation capable

of operating in times of crisis independently of any future
closures of the Suez Canal but also to encourage oil companies to explore and develop alternate sources of petroleum
supplies in other parts of the world.  The contemporary
global movement toward the diversification of Western Europe's sources of energy had thus been initiated by Eisenhower under the impact made upon him by the Suez Canal petroleum crisis of 1956-57 and the resultant greater Soviet
penetration into the region.


## SUMMARY AND CONCLUSIONS

The great importance for U.S. foreign policy of a continuous flow of oil from the Middle East to Western Europe
was clearly demonstrated in a successive series of oil emergencies, beginning with the Marshall Plan, running through
the Iranian oil crisis, and, finally, ending with the Suez
Canal stoppage of 1956-57.  In order to secure that Middle
East oil for the OEEC nations and NATO, the United States
spent billions of dollars, organized a tremendous worldwide
administrative machinery, negotiated with multinational oil
corporations, set up ocean lines of tanker fleets, and even
risked charges of knowingly violating its own antitrust laws.
In all that gigantic effort to keep Middle East oil
moving west, the United States had only a very minimal economic interest.  Since Eisenhower, in 1959, had limited the
importation of Middle East oil to 2.5 percent of general
U.S. consumption, the U.S. economy became nearly completely
independent of this source of oil supply.  Apart from its
significance to the prosperity and security of U.S. allies,
Middle East oil had no economic or political lever on the
U.S. market.  In contrast, however, the Middle East region
itself was becoming increasingly important to the U.S. policy of the containment of communism.  The concern about the
emergence of the Soviet Union as a major Middle Eastern
power superseded, on the U.S. scale of priorities, all past
oil policies.
At the end of World War II, Soviet involvement in Middle East affairs was primarily confined to the northern-tier
nations.  Soviet pressures were made at that time on Greece,
Turkey, and Iran and, in turn, produced the Truman Doctrine
of 1947, which enunciated the policy of containment.  But
after the defeat of Mossadegh's nationalist regime and the
return to public disfavor of the communist Tudeh Party in
Iran, Soviet policy seemed to slacken its pressures on the
northern-tier nations and began, instead, to focus its

attention on the Arab countries of the fertile crescent and
Egypt.  The new friendships were cemented by arms deals,
military missions, and loans.  Soviet prestige in the region
was on the increase.

The Suez Canal crisis offered the Soviet Union an un-
paralleled opportunity to reinforce its great-power status
among the Arab nations of the Middle East.  With great pub-
licity and fanfare, the Soviet government came to challenge
the Western powers' attitude to the nationalization of the
Suez Canal Maritime Company as being based on relationships
that had been "brought about in the past through conquest
and occupation" and that were therefore "no longer proper."
Accordingly, the Soviet government considered the "Egyptian
Government's decision to nationalize the Suez Canal Company
as a perfectly lawful action following from Egypt's sovereign
rights."[80]

At the London Conference on the Suez Canal, however,
Soviet Foreign Minister D. T. Shepilov seemed to be address-
ing the new nations of Asia and Africa more than he was trying to convince the Western delegates of their mistaken
military path.  Thus, he said during that conference, these
military measures would be "given a fitting rebuff not only
by Egypt but also by other peoples fighting for their sovereignty and national independence."[81]  The Soviet government's statement on the Suez Canal Conference also addressed
itself to the Third World nations, stating that the planned
military expedition would "undoubtedly . . . arouse the pro-
found indignation of the peoples of Asia and Africa against
the Governments of the countries that were embarking on the
path of aggression."[82]  And again, the Soviet statement asserted that "the threats to use force in relation to Egypt
are being decisively condemned by the public all over the
world."[83]

The appeal to the Third World that came out so clearly
in the Soviet government's statements and Shepilov's arguments at the London Conference did not escape the attention
of the U.S. secretary of State.  It very quickly forced a
policy turn in the Department of State.  The initial support
of the notion of a military expedition to the Suez Canal that
Sir Anthony Eden believed he was getting from Washington now
changed quickly into a series of legalistic declarations
that duplicated, to some degree, the Soviet definition of
the "two aspects" of the question of the nationalization of
the Suez Canal.  In accordance with this legalistic definition of the situation, the Dulles scheme for a Canal Users'
Association was designed to pacify the new nations, block
any further Soviet denunciations, and use a legal formula,

rather than a military action, to secure Western demands. According to Paul Johnson, "After a few months of this, Mr. Dulles assumed, Nasser would be prepared to compromise."84

When the military expedition on Suez actually began on November 5, the U.S. government, to its great dismay, found itself impelled to pick up the tab--the penalty for its unfinished policy. Once again the U.S. people assumed upon themselves the responsibility for Europe's economic survival. Since the Marshall Plan, each Middle East crisis had placed an additional burden upon the United States in its determination to preserve the prosperity and security of that continent. At the same time, however, each new crisis in the Middle East had spurred European governments to encourage their industries to supplement foreign dependence with domestic production and eventually phase out the former altogether. Gradually, as a result of balance-of-payments considerations, a prosperous European petrochemical industry successfully replaced dependence upon Arab refineries. Moreover, the Iranian oil crisis pointed out clearly the continent's dependence upon the major Anglo-American-Dutch oil corporations.

To avoid this uncertain dependence, the continental governments began to encourage their own national oil companies to go into foreign oil explorations. These early beginnings of European involvement in Middle East oil developments received a tremendous push from the Suez Canal petroleum crisis. Continental governments agreed with Eisenhower in principle on the need to diversify sources of oil supply in order to avoid future crises. But their policy of "diversification" also emphasized a diversification of suppliers. It no longer seemed to them justifiable, either economically or politically, that they should buy all their crude from the same major Anglo-American-Dutch companies that had been dominant for so many years in the field of oil. The national oil companies of the continental countries entered competition for Middle East oil by offering the producing countries far better terms than the cartel of the seven ever dreamed of conceding. The producing nations, on their part, welcomed these agreements not only for their economic profit but also for the opportunity the agreements offered to rid themselves of an exclusive dependence for their revenues on a limited number of multinational corporations.

Thus, the Suez crisis can be seen as the end of one era and the beginning of another in more than one way. On the one hand, the U.S. scale of policy priorities in the

Middle East was fast adjusting to meet the Soviet challenge in the region.  This rivalry was a far cry from the commercial competition with Great Britain for oil concessions during the 1920s or even the attempts that had been made by the U.S. government to enter Middle East oil developments during the mid-1940s.  The Soviet presence in the Middle East now posed new problems and raised national issues in areas of military strategy and political relations with other nations.  In short, the Soviet threat in the Middle East typically replaced economic considerations by the overriding concept of survival, thus once again confirming the importance of the latter over all other issues or objectives. The U.S. oil corporations operating in the Middle East could no longer look to Washington for protection, since the new realities of that region could impose upon the State Department a policy that, at times, might run contrary to the interests of those companies.

On the other hand, new drives were brought to bear upon the traditional structure of the international oil industry with the emergence of new European and independent U.S. oil companies as worthy competitors.  This change in the economic structure of the oil industry abroad put into motion two sets of developments:  First, Europe became steadily more self-sufficient as a supplier of its oil requirements than it had ever been before.  Second, the large oil corporations, for a variety of reasons, were converting into multinational structures, thus adding to their credit many national bases in the consuming countries as well as the producing countries.  These two trends released the United States from the responsibility for European energy supply. As subsequent events proved, the Suez crisis of 1956-57 was in fact the last instance when the United States was called upon to save the European economy from sinking due to oil crises in the Middle East.  In 1967, when the canal was again closed to all sea-borne traffic, U.S. aid, although offered in the same manner as in 1956, was not required.

# III

## PEACEFUL COEXISTENCE
## WITH ECONOMIC COMPETITION:
## THE PETROLEUM PICTURE

# 8

## SOVIET PETROLEUM
## OFFENSIVE

The Khruschchev era offered U.S. decision-makers an example of what the Soviet Union can do with its immense oil resources in order to gain both economic and political advantages abroad. To those Americans engaged in oil enterprises, the Soviet profession of peaceful coexistence inspired the specter of a strangulating economic competition between East and West in a vast number of foreign markets. During the period between the Suez crisis of 1956–57 and the fall of Nikita Khrushchev from power in 1964, the Soviet Union rivaled Western oil trade everywhere in the noncommunist world, while using communist countries as an exclusive closed market. This foreign oil policy came to an end with the change of regimes in the Soviet Union and the return to more traditional economic principles. Whatever foreign oil contracts and commitments to help build refineries or construct pipelines existed after 1964, they were largely agreements entered into under Khrushchev and awaiting their fulfillment or date of expiration. By the time that the Suez Canal was again closed to all sea-borne transportation in 1967, there were only very few such agreements still in existence.

The first section of this chapter deals with the U.S. image of Soviet ideology and its relation to policy. Such interpretive images were important inasmuch as they came to be a motivating force in shaping the direction of U.S. responses to Soviet actions. The second section describes the conversion of the Soviet economy from autarkianism to internationalism and attempts to relate this change to the evolution of doctrine as seen in the writings of the economist Eugene Vargas, Stalin's book Problems of Socialism, and the 19th Congress of the Communist Party of the Soviet Union (CPSU). The third section analyzes the Soviet oil offensive under Khrushchev, its <u>raison d'être</u>, its accomplishments for the Soviet Union, and the threat it posed to the West. Western response to this Soviet economic and political threat, such as it was, is dealt with in Chapter 9.

## THE U.S. IMAGE OF THE SOVIET UNION
## IN THE POSTWAR YEARS

The Cold War gave rise to a new breed of scholars who very soon came to be known as Kremlinologists. They were social scientists who studied every minute occurrence related to the Soviet Union and every nuance of any given aspect of Soviet life. Due to their professional preoccupation with detail, they were often afflicted with varying degrees of myopia. Among their several obsessions, there was a great concern about the study and interpretation of all communist "scriptures," in the process of which many irrelevant utterances were magnified out of all proportion and were given an undeservedly prominent status. There did not seem to be, at first, any quotes from any Soviet sources that were ever assigned a secondary place of importance. All references seemed to be awarded equal significance in the overall communist doctrine, and, therefore, all were equally disturbing and puzzling inasmuch as, as was often pointed out, there were substantial contradictions among so many of them. Consequently, there was an immense effort spent by many Kremlinologists to reconcile these contradictions in order to clarify what was assumed to be a unified and cohesive single communist theory. Since such a unity never existed in reality in communist thought, such an effort naturally resulted in a great deal of fruitless sophistry.

U.S. Kremlinologists were even more sold on the idea of the unity of communist doctrine than were their Western European colleagues. As a result, when the United States became after World War II the leading nation in the Western world and its new world view had assigned the Soviet Union the role of the designated enemy of the country, the theory and practice of communism became of increasing concern. Within this area, the communist doctrine on world revolution seemed to be of central interest. Millions of words poured from many a Kremlinologist's pen describing the Marxist view of the rising Western proletariats and the assumption by their leaders of an elitist dictatorship on behalf of the working masses. However, the emergence in the 1950s of new nations in the former colonies of Asia and Africa and the predicted rise of a Third World political force gave a new stimulus to the avid reading of Lenin and Stalin in order to detect in their writings whatever they might have said in regard to the colonial peoples and their function in the world revolution. Lenin's and Stalin's speeches, writings, and conversations were sifted thoroughly, and an organized field of study was soon developed that concerned itself with this line of research. It was once again remembered that when the revolutionary zeal in Europe, upon which Lenin seemed to depend so heavily during World War I years, was beginning to recede, he changed his concept of the course of world revolution and switched his interest to the Far Eastern nations. Accordingly, he called for the awakening of all colonial peoples to throw off the imperialist yoke and to develop rapidly the revolutionary movement.[1] Not only did Lenin want these colonial peoples to attain their freedom from the imperialist nations, but he also set out to provide them with a guide and an inspiration by setting before them the example of the Russian revolution, thus claiming for the Soviet Union the leading role in the world revolutionary struggle. Nevertheless, in reality, the Eastern nations offered Lenin only a temporary compensation for the loss of the European proletariat. Lenin, still a Marxist, continued to believe that the true world revolution must be the work of the working masses in the industrialized centers of the world. The preliminary step to this revolution, however, must be taken in the West's source of supply and exploitation, the colonial countries. "Genuine Communism," Lenin informed a Japanese interviewer, "can thus far succeed only in the West. However, the West lives on account of the East. European imperialist powers support themselves mainly from Eastern colonies. But at the same time they are arming their colonies and teaching them to fight. Thereby the West is digging its grave in the East."[2]

Lenin then conceded that temporary agreements and blocs could be made with bourgeois liberation movements in colonial countries. The Communist International, he said, "must join in a temporary alliance with the bourgeois democrats of the colonies and backwards countries, but not merge with them, and must unconditionally preserve the independence of the proletarian movement even if it is in a quite incipient form."[3]

This view was endorsed by the Second Congress of the Communist International, which also explicitly stated that

"the first step of a revolution in the colonies must be to
overthrow foreign capitalism."

Stalin's approach to the problem of the revolutionary
movement in colonial countries was in essence very similar
to that of Lenin and the Communist International.  He dif-
ferentiated between a revolution made by oppressed workers
against their bourgeois exploiters, which would be strictly
a class struggle, and a revolution by a colonial nation
against the foreign occupiers of its homeland.  In the
colonial countries, then,

> the oppression exercised by the imperialism
> of other states is one of the factors of
> revolution; this oppression cannot but af-
> fect the national bourgeois also; the na-
> tional bourgeois, at a certain stage and for
> a certain period, may support the revolution-
> ary movement of its country against imperial-
> ism, and the national element, as an element
> in the struggle for emancipation, is a revo-
> lutionary factor.[4]

Although a research into communist political thought
was necessary in order to understand the theoretical molds
of Soviet leadership, it also had its limitations inasmuch
as an overemphasis on doctrine led many scholars and poli-
ticians to overlook some objective realities of international politics.  Traditional Russian territorial convul-
sions and contractions, representing an age-old attempt to
break away from the strictures of its own land-bound national geography, were interpreted in the United States as
crusades motivated by a revolutionary doctrine, while Russia's traditional strategic need to secure all its flanks
by means of a deep belt of friendly buffer states was considered to be a consequence of the dogmatic utterances made
by the gentlemen in the Kremlin.

In reality the Soviet Union did not lose sight of its
historic goals.  The experience of World War II not only
did not dim these needs but in fact reinforced them inasmuch as the Soviet leaders had acquired a taste for global
Realpolitik.  Emerging from that war as the second great
power in the world enabled the Soviet Union to abandon its
defensive prewar posture and to formulate instead a forward
policy calculated on regional subtleties of balance-of-
power relationships while keeping the other superpower at
bay upon a seesaw in a precarious balance of terror.  Indeed, this strongly self-assertive foreign policy was

greatly influenced by the Soviet world view. In Soviet eyes
the future of the world was going to be determined "by the
outcome of the trial of strength between the two basic systems of politico-economic organization--the capitalist and
Communist. The struggle for hegemony between these two systems as Soviet strategists see it, is the vital contest of
our age."[5]

U.S. observers thus continued to attribute great significance to the Kremlin's doctrinal calculations within the
Cold War concepts.[6] But several political and military actions of the Soviet Union during that period did not seem to
fit into the general scheme of what was in effect a U.S.
theory of communist ideology. Soviet policy in the Suez
conflict, even more than its armed violation of Hungary's
territorial sovereignty, appeared to have destroyed much of
this carefully assembled doctrinal structure and to cause a
reevaluation of Soviet motivations. Some facts of recent
chapters in history were now brought down from the dusty
shelves and reexamined. Molotov's communiqué to Hitler of
1940, in which he advised the Nazi dictator that the Soviet
Union's area of aspirations "lay south of the Baku-Batum
line in the general direction of the Persian Gulf," it was
observed, contained no doctrinal or revolutionary message.
Soviet claims for a share in the trusteeship of the former
Italian colonies did not seem to be motivated by any considerations of the future of the local proletarian class, which,
in fact, did not exist in those areas. Nor was the Soviet
Union's abrupt reversal of itself and its support of the
Italian claim for a trusteeship of its own former colonies
intended to be more than a Realpolitik playoff on an international scale. It was equally observed that Soviet tensions with the governments of Turkey and Iran were also political and bore no direct relation to any communist tenet.
True, in order to further its ambitions the Soviet Union
did have a weapon that was not available to such a great extent to Czarist Russia. While the Russian Orthodox Church
abroad could be relied upon to support all Russian aspirations, it had never filled the enormous role that the foreign communist parties constituted as an arm of Soviet foreign policy. In this realm, a special role was assigned
after 1952 to the communist parties in the developing areas.
The part of the growing intelligentsia of the former colonial nations that was attracted by the Soviet revolutionary
example consistently supported the Soviet line. Being the
most articulate segment of the largely illiterate population
of these countries, it seemed able to excite public opinion
and occasionally influence political events. Soviet reluctance to support the indigenous communists at the expense of

its own relations with the existing governments again confirmed Western observers in their belief that doctrine and
policy were often kept very much apart in Soviet practice.

Once the ideological contents of the Soviet regime and
its contribution to the formulation of foreign-policy-making
was set in its proper perspective, U.S. government departments as well as U.S. scholars began to concern themselves
with the military, political, and economic aspects of the
Cold War. This approach was novel inasmuch as it had
ceased to subordinate military and political considerations
to ideology. It made a closer examination of economic policies, as both military and political strengths were basically dependent upon economic strength. Thus economic assessments of the opponent became paramount in policy formulation
in both blocs. Not only were communist dogmatists looking
forward to the West's economic collapse as a sign of its
overall deterioration, but the West itself, in reverse conditioning, was anxiously examining growing Soviet affluence
in international trade as a sign anticipating a proportion-
ate growth of Soviet political influence.


SOVIET FOREIGN TRADE


In Stalin's time, in the early 1950s, a growing number
of Soviet economists began to advocate the view that the
Soviet economy "could derive substantial advantages from an
expansion of its international trade."[7] Economic considerations as well as political evaluation of the opponent's
world strategy compelled the Soviet Union to reorient its
production efforts toward development and expansion of its
foreign trade. The Soviet Union seemed set on reaching the
U.S. level of commercial and military production and foreign trade. In order to compete commercially with U.S. exports, both for political and strategic reasons as well as
to gain hard currency exchange, the Soviet Union had to reverse its traditional foreign trade policy. Instead of
concentrating on the import of heavy machinery to create an
autarkian economy, as had been done under the Stalin regime
when trade was subordinated to the overriding goal of self-
sufficiency, the Soviet Union began in the 1950s to turn
the flow of trade outward, trying to reach, in the process,
as many nations as possible. However, as before, Soviet
exports continued to consist of raw and semifinished materials, sold in bulk. Thus, the second most powerful country in the world was, in essence, repeating an export pattern similar to that of any underdeveloped nation. Its

most exportable products, under this system, were wood, manganese, aluminum, petroleum, and petroleum products. Despite the ratio increase over the low level of Stalin's time, Soviet exports to the free world in 1959 were still only at the level of a country the size of Denmark.[8]

The limiting factors responsible for this trade limitation were sowed during the Lenin and Stalin eras, when Soviet policy deliberately strove for self-sufficiency and, consequently, was developed in isolation from foreign trade. Soon after the October Revolution, trade was nationalized by the Soviet government not in order to expand outward and participate and compete in the world economy but in order to build a self-sufficient economy, protected from strangulation by capitalist monopolies and hostile foreign governments. Thus, a decree dated April 23, 1918 prohibited any import and export except for those conducted by government agencies. This decree aimed at protecting Soviet economy from the hostile outside world. It strove to achieve the isolation, and thereby the independence, of the ruble from the influence of foreign exchange markets.

While succeeding in isolating the Soviet economy from the rest of the world, this system of centralized government control over all foreign trade turned into a very powerful weapon in the hands of the Soviet government once the government decided to expand its trade and compete with the West for international markets. In this effort, the Soviet government had certain advantages over the West: For example, it was in a strong bargaining position in relation to noncommunist governments because it could discriminate between purchasers without having to give undue regard to the normal balance-sheet considerations of gain-and-loss imperatives that motivate free enterprise decisions. This power enabled it to use its leverage in foreign trade in order to further political and strategic objectives.

Under Khrushchev, foreign trade assumed new dimensions. It took a sharp turn toward a rapid penetration of all international markets. From the start, this so-called Soviet economic offensive was politically oriented. Not only did the Soviet Union seek economic gains, but it also used its foreign trade as a tool by which to attain political and military objectives. Since trade was centrally controlled by the government, it could be manipulated and diversified in a manner that supplemented and increased the Soviet bloc's potential. Trade was thus utilized to attain considerable growth of Soviet influence in the economic and, therefore, also the political life of other, noncommunist nations.[9] As an economic tool, trade was used in order to

obtain hard currency where possible and, failing that, to
obtain by barter vital materials (see Table 15, below) and
technical know-how from the industrialized nations, which
seemed more than willing to barter short-run Western economic profits for long-term Soviet advantages.  It concentrated on the buying of strategic goods produced by the
more advanced Western technology, especially the import of
such plants as could be duplicated by its own scientists.
Among the barter items that the Soviet Union showed marked
preference for were

> petrochemical and synthetic plants, elec-
> tronic equipment for communications and con-
> trol, precision and highly automatic machine
> tools, construction machinery, industrial
> handling equipment, carbon steel and alloy
> sheet and strip, modern cold-rolling mills
> for transmission equipment, precision bear-
> ings, rail and ocean transport equipment,
> complete tire plants, and large diameter
> pipe and other equipment needed for the pro-
> duction and transport of oil.[10]

As a political tool, foreign trade was used in the
former colonial areas to expand Soviet influence.  This
was done in a variety of ways:  for example, by offering
cut rates and accepting in exchange agricultural commodities or local raw materials or, even, soft currencies.  In
this way, the Soviet Union not only undermined the position
of Western competitors but also tied the economy of the
client nations into dependence upon Soviet markets for dis-
posing of their surplus commodities and, in turn, relying
on Soviet sources of supply.  Achieving this dependence,
some of these nations found out that political strings were
attached to the economic largess.  It posed a significant
threat to the stability of their economy inasmuch as Soviet
trade with them could be abruptly interrupted at any time
for political reasons that suited the Soviet bloc, and regaining access to these closed markets once again might involve considerable concessions on the part of the clients.
A study committee of the National Petroleum Council concluded that

> Any substantial reliance by a Free World
> country on trade with the Soviet Bloc
> the Government of which exercised complete
> control over foreign trade, creates a

> threat to the security, political indepen-
> dence and economic health of that country.
>     Soviet Bloc markets can be . . . and
> have been . . . closed and Soviet Bloc sup-
> plies can be . . . and have been . . . in-
> terrupted more for political reasons than
> for commercial considerations.[11]

Examples of such interruptions of normal trade relations
for political reasons were those with Israel in 1956, Finland in 1958, and Poland in 1963.

Its large margin of maneuverability made it possible
for the Soviet Union to reach an increasingly large number
of nations and to make its presence felt in far-flung parts
of the world.  However, "despite this increased trade," concluded the National Petroleum Council's study committee,
"it would be unrealistic to expect the U.S.S.R. trade to become permanently interwoven with that of the Free World."[12]

In other words, in spite of its enlarged international
trade, the Soviet Union was still viewing the outside world
as hostile to it.  It aimed to maintain the independence
and isolation of its economy by keeping it apart from the
influences of foreign trading and exchange markets.  At the
basis of its world view, then, the Soviet leadership still
believed in the essential contradiction between capitalist
and socialist economies.  This contradiction could only be
resolved by struggle between the two systems.  It would be
dangerous, therefore, for the Soviet Union to make its own
economy, by means of foreign trade or otherwise, dependent
to any extent on the opponent's economic systems.  The future of the world, then, in the Soviet world view, was to
be determined, according to a Senate committee report, by
the "trial of strength between the two basic systems of
politico-economic organization--the capitalist and the Communist.  The struggle for hegemony between these two systems, as Soviet strategists saw it, was the vital contest
of the age."[13]

The turning point in the relations between the Soviet
Union and the other bloc countries on the one hand and the
underdeveloped countries on the other came with the development of a new political line that simultaneously fitted
into the "peaceful coexistence" strategy.  The first requirement was a change in the policy followed by the communist parties in the underdeveloped countries.  In response to the dictate of the Soviet Union, these parties
altered their tactics: The final objective, seizure of
power, was no longer to be achieved by force and by acting

publicly against the official policy of the underdeveloped
country concerned, but rather by "peaceful political action
and the building up of communist fronts with the forces of
nationalism."[14]
     Thus, in order to continue to maintain good political
as well as trade relations with the new nations of Asia and
Africa, the Soviet Union had to come to terms with bourgeois, often anticommunist governments.  It had to bolster
these diplomatic relations by gestures of technical and economic assistance out of all proportion to its own economy.
Fearful that Western policies might lead the new nations of
Asia and Africa into an anti-Soviet coalition, the Soviet
Union radically changed its own policy toward these countries.  The Soviet Union's response to the political and
military forces that seemed to threaten it was to adopt a
program that emulated U.S. tactics primarily in the attempt
to acquire influence in the underdeveloped world.
     Petroleum, which existed in great abundance in the
Soviet Union, became the most essential single product of
Soviet export as well as the major tool of Soviet economic
offensive under Khrushchev.  (See Table 15, below.)  The
increased rate of exploration and drilling during the
Khrushchev period far exceeded the demand of Western European consumption from Soviet sources and thus created large
surpluses.  The surpluses were dumped on the underdeveloped
world for political and strategic gains.


THE SOVIET PETROLEUM OFFENSIVE


     The great havoc visited upon the Soviet petroleum in-
dustry during World War II necessitated an immediate reorientation for the intensification of exploration and drilling.  Instead of venturing into experimental and as yet un-
proved new areas, efforts were concentrated in the postwar
period on proved priority regions.
     By the mid-1950s the Soviet Union had reached a large
scale of oil production through the development of the
Bugulma-Beleby Basin, which was known to hold half of the
nation's 10 to 11 billion metric tons of industrial petroleum reserves.
     The expansion of the petroleum industry was a dominant
factor in the urbanization and industrialization of the
Soviet Union.  Theoretically, it permitted Khrushchev to
shift priorities and redirect certain resources into various consumer goods industries.  In reality, however, the
resources were largely invested instead in a rapid buildup

of powerful military industries. This trend signified the failure of Khrushchev's efforts, as reallocations of resources still continued to minimize the role of consumer goods and maximize the role of heavy and military industry products and exports.

Soviet development of its petroleum industry under Khrushchev, then, represented a tremendous economic achievement that made itself felt not only in the domestic heavy industry of the country but also in the nation's increasing role as a powerful factor in the world market. As a result of the remarkable expansion achieved in drilling and refining, the Soviet Union once again became a net exporter of oil and oil products.* By 1955, the year in which Khrushchev came to power, the intensified efforts expended in the Soviet oil industry during the prior 10 years had recovered for the Soviet Union its position in the world oil market to about one-half of the 1929 level. By 1959, its position was 5.6 times that of the 1929 level.[15]

Khrushchev's accession to power became a turning point in the development of the Soviet oil industry, whose character now changed from primarily a domestic-oriented industry into a tool of foreign policy, a trump card in the world arena, influencing the character of international relations. Under Khrushchev, oil became the major weapon of the Soviet foreign trade offensive. It became a tool to rival and combat the Western oil companies' operations, to which Soviet doctrine had always attached a decisive role in the U.S. policy-making processes, and a means by which weak national economies were made to become dependent upon Soviet source of supply.

Soviet oil exports during the 1950s show the year 1955 as a turning point:

---

*The traditional Soviet position in the world market as a net exporter of petroleum and petroleum products had only been interrupted by World War II, at which time it became a net importer of fuel. In 1944 alone, under lend-lease provisions, the Soviet Union received from the United States 5.15 million barrels of aviation gasoline, 266 pounds of bromide for fuel additive, and 6.8 million pounds of grease. (See Demitri B. Shimkin, Minerals: A Key to Soviet Power [Cambridge: Harvard University Press, 1953], p. 66.)

| 1950 | 265 million barrels |
| 1955 | 500 million barrels |
| 1958 | 1 billion barrels |

Thus, in 1958 total Soviet petroleum production approached that of Venezuela, the world's second largest oil producer, and amounted to about one-third of the annual production of the United States. Within two years, in 1960, Soviet production surpassed that of Venezuela, and became more competitive with U.S. oil production, whose growth rate was much lower.[16]

However, not only were political motivations involved in this Soviet foreign policy, but there was also a strong economic factor: An undeniable need for foreign exchange gave a great incentive to this penetration of Western oil markets, as a result of which the Soviet Union became one of the largest sources of petroleum supplies to these countries.

### Soviet Regional Petroleum Markets

In the period between the two world wars the European countries of the Mediterranean region had purchased nearly half of all Soviet oil exports. The remaining portion of the Soviet foreign oil trade had been divided about equally between nations of Western Europe on one hand and the Far East and the Western Hemisphere on the other (see Table 14).

### TABLE 14

### Soviet Prewar Foreign Petroleum Markets

| Mediterranean Region | Western Europe | Far East and Others |
|---|---|---|
| Italy | Germany | Japan, mainly; |
| France | Great Britain | sporadically: |
| Spain | Belgium | United States |
| Greece | Sweden | India |
| Egypt | Denmark | Others |
| 50 percent | 25 percent | 25 percent |

Source: Demitri B. Shimkin, Minerals: A Key to Soviet Power (Cambridge: Harvard University Press, 1953), p. 66.

When the Soviet Union, during the 1950s, had once again started to penetrate foreign markets with large quantities of petroleum supplies, its first goal was to regain its prewar position among its former client nations.

The Mediterranean Region. In trying to recapture its prewar markets the Soviet Union directed its attention, first, to its traditionally most lucrative clientele, namely, the Mediterranean nations. The Soviet press described this ordinary international business competition for foreign markets as being a "rightful claim" for "traditional markets," and, based upon this view of history, the Soviet Union was in fact the primary "legitimate" supplier to the Mediterranean countries. But the "historical" role of Soviet petroleum in this area was blown out of all proportion to its actual prewar status, when energy outside the United States was still derived mainly from coal. The Soviet reentry into these markets was thus justified not as an aggressive commercial venture such as may be ascribed to the imperialistic nations but as a "right" that was backed by the "legitimacy" of history.

But in returning to the Mediterranean, the Soviet Union encounted a new political complex. Cold War stresses had molded new trade patterns. While the 1955 total of petroleum and petroleum products exports to the Mediterranean countries equaled 75 percent of the Soviet 1934 total, the market composition itself had, in effect, undergone a significant change. Shipments to NATO nations did not resume their prewar level: Exports to France fell by 25 percent, to Greece by 50 percent, and to Italy by 66 percent, and Spain, a nonmember, no longer ordered fuel from the Soviet Union. On the other hand, Egypt doubled its orders, while Israel and Yugoslavia became active new customers. Also during this time, but for a short time only, Albania became a moderately important supplier to the Soviet Union.

The year 1955 proved to be a turning point in Soviet petroleum trade. Very shortly after Khrushchev's accession to power, the trend of Soviet petroleum trade with countries of this area was again reversed. In 1959, net Soviet exports to Mediterranean nations rose three times above the 1934 level and four times above the 1955 level. The most remarkable change occurred in Italy, where shipments exceeded the 1934 level by 164 percent and were eight times as great as in 1955 and provided at that time 17 percent of the Italian total consumption of petroleum. Next in importance in this region were the markets of Egypt and Syria,*

---

*Most of the exports to Egypt and Syria were reexported.

169

which, together, purchased 13 times more fuel from the Soviet Union than in 1934. French imports of Soviet petroleum increased by 25 percent over the 1934 level. On the
other hand, Greece, where Soviet oil supplied 21 percent
of total consumption, was in reality a minor customer. At
the same time, shipments to Israel had been stopped for political reasons, and the Yugoslavian market, which in 1959
had been cultivated to twice its 1955 capacity, was soon to
be lost as well. Similarly, Albania's role as a supplier
was also short-lived.

The Western European Region. If the Mediterranean region, in the Soviet outlook, constituted a traditional market for its oil exports, the rest of Western Europe was a
relatively recent major customer.

In 1955, at the beginning of the Soviet oil offensive
in Europe, Soviet petroleum exports to these countries rose
by 140 percent above the 1934 level. By 1959 they again
rose by another 140 percent above the 1955 level. This
1955-59 expansion represented a sixfold increase for that
period. The major part of these exports was shipped to
Sweden, Finland, and Iceland, and the last two bought their
entire petroleum product imports from the Soviet Union. In
Sweden, where petroleum consumption was much higher than in
the other two countries, the ratio of Soviet exports accounted only for 14 percent of the total.[17]

The Far Eastern Region. In 1959 Soviet oil exports to
countries in the Far East reached a peak of 5.4 million
metric tons as compared with 630,000 metric tons in 1934.
This expansion was accounted for by the rising needs of the
People's Republic of China, North Korea, Outer Mongolia,
and North Vietnam as well as the restoration of trade with
Japan. But, by that time, successful oil drillings in
China had reduced that country's dependence on Soviet oil,
the importing of which in 1960 was cut by 45 percent.[18]


Oil for Technology


Plans for the steady increase of petroleum exports in
the future were prepared and based on calculations of similar increase in Soviet need for foreign exchange, which
would be invested in the purchase of chemical and fertilizer
plants as well as heavy machinery for its industry. (See
Table 15.) The need to increase petroleum exports, then,
was reasonably anticipated by the most pressing Soviet needs
to fulfill blueprinted plans. It was imperative for the
Soviet Union that its oil reserves and scheduled drilling
be adequate to support its planned production effort.

TABLE 15

Commodity Composition of Soviet Foreign Trade
with Industrial West

|  | 1960 | 1962 | 1964 | 1966 |
|---|---|---|---|---|
| **Exports (percent of total)** | | | | |
| Fuels and lubricants | 26 | 29 | 31 | 27 |
| Ferrous and nonferrous metals | 11 | 11 | 15 | 14 |
| Wood and wood products | 16 | 18 | 21 | 17 |
| Foodstuffs | 12 | 12 | 6 | 7 |
| Furs and pelts | 5 | 4 | 4 | 4 |
| Other exports | 30 | 26 | 23 | 31 |
| Total exports | 100 | 100 | 100 | 100 |
|  | | | | |
| **Imports (percent of total)** | | | | |
| Machinery and equipment | 43 | 47 | 36 | 32 |
| Ferrous and nonferrous metals | 28 | 23 | 4 | 5 |
| Wool and synthetic fibers | 6 | 5 | 4 | 4 |
| Wheat and wheat flour | -- | -- | 31 | 24 |
| Other imports | 23 | 25 | 25 | 35 |
| Total imports | 100 | 100 | 100 | 100 |

Source: U.S. Congress, Joint Economic Committee, Subcommittee on Foreign Economic Policy, Soviet Economic Performance 1966-67, 90th Cong., 2d sess. (Washington, D.C.: Government Printing Office, 1968), p. 104.

However, the inevitable effect of this expansion of oil drilling resulted in an accrued need to purchase more Western oil industry equipment and technology in order to increase the Soviet ability to produce, refine, and export oil. This spiraling circle caused the Soviet Union to spend its earned foreign exchange on the expansion of its petroleum production for Western European consumption instead of redirecting it, as had originally been intended, toward the expansion of domestic industries, agriculture, and consumer goods.

Thus, crude oil production in the Soviet Union had increased rapidly and soon exceeded official plans without, in effect, contributing the desired advantage that had set up the objective of this foreign trade in the first place. It is quite significant that of all the forms of primary energy in the Soviet Union, only crude oil production exceeded its planned goals, and at the same time its financial earnings seemed to have been reinvested in its own expansion while proportionately far less than desired of these totals was redirected into the development of other industries.

Soviet Price Policy

Soviet price policy for petroleum supplies was dictated by overall Soviet goals, chief among which was the overriding objective to win in the economic competition with the industrialized West. Immediate gains were thus sacrificed in order to attain this long-range objective. The following are the main Soviet methods:

1. Cut Rates: The price for Soviet oil and oil products was marked well below the standard generally accepted in the international market and with no demonstrated relevance to the exorbitant prices charged for oil in communist states, which were, in fact, located geographically closer to the source of supply. Thus, in 1957 the Soviet Union's average price in the international market was $2.06 per barrel, as compared with $2.92 charged by Venezuela and $2.79 charged by the Middle East petroleum companies. These cut rates continued to be progressively decreased over the years. In 1958 the Soviet Union charged Argentina only $1.60 per barrel, thus undercutting Venezuela's world oil average price by $1.32 per barrel. At the same time the Soviet Union was selling oil to Poland for $2.87 per barrel.[19] Evidently, normal business considerations could not have motivated this price policy and other, noneconomic, reasons must have prevailed. The Soviet Union reached a rock-bottom figure when, in 1959, it started to sell petroleum to Italy for $1 per barrel.[20] Italy was then, as it is now, the most important NATO member in the Mediterranean region.

2. Foreign Refining Facilities: Soviet petroleum contracts with other nations called for business deals that were generally beyond the reach of private enterprise companies. The Soviet Union's contracts insisted on the use of facilities that originally had been built, at great expense, by the international petroleum industry. In Cuba this practice became possible when the Cuban government nationalized all foreign petroleum facilities in the country. In India, the international petroleum companies refused to accept Soviet petroleum for refining, as a result of which the Soviet Union assisted the Indian government in the construction of state-owned refineries. The prospects of such refining facilities, which would be in a

position to supply not only the whole Indian subcontinent but also important foreign markets located in Japan, Australia, and South and East Africa, were sufficient to dim a threat issued by the Standard Oil Company of New Jersey to blacklist all tanker operators carrying such Soviet oil from India. The current state of unemployment in the tanker shipping lines was such that enough tanker owners were only too glad to put their idle tankers to any use.

3. <u>Barter</u>: In the absence of cash payments, the Soviet Union showed a willingness to trade its petroleum in exchange for local commodities. In the industrialized areas of the world, these barter payments took the form of heavy equipment for the growing Soviet petroleum industry. Soviet salesmen offered attractive prices for their oil in exchange for pipelines and refineries from Western nations. The Soviet Union used this equipment to enlarge its oil fields, build many more refineries, and construct a vast network of pipelines for the transportation of petroleum, crude and products, as well as natural gas to all territories under its domain, including nations of Eastern Europe and docking ports on the Baltic and Black Seas.

4. <u>Soft Currencies and Indigenous Commodities</u>: The Soviet Union further diversified its price policy to suit the needs of the weaker economies of the developing nations. While the international petroleum companies were selling oil in return for hard currency, the Soviet government entered into deals with governments of Asia, Africa, and Latin America, according to which oil was exchanged for "soft" local currencies that were valueless on the international market. The Soviet government also offered to trade oil in exchange for local commodities such as sugar, cotton, fish, and coffee. Obviously, barter agreements of this kind were beyond the range of the Western oil companies, which had "to obtain profits in terms of convertible currencies in order to stay in business."[21]

Soviet flexibility in setting its prices and barter agreements to suit the desirability of each customer was a result of three basic factors:

1. Soviet foreign trade of all commodities, including that of petroleum, is a state monopoly and thus it can enjoy a high degree of budgetary maneuverability.

2. The Soviet oil industry is the largest integrated oil enterprise in the world; uncomplicated by intricacies of cartel agreements, it is centrally controlled and directed.

3. Soviet state ownership of petroleum resources, refining, transportation, and distribution offers its government a tool of a high political and strategic potency, which may influence economic decisions.

Thus, competing with Soviet oil trade was not an easy task for the Western companies, especially where Soviet moves were politically motivated. Most of the Soviet sales in the world markets were made on a government-to-government basis, where official contracts could and did often include references to noneconomic matters. Thus, oil transactions seemed to be made incidentally to the avowed purpose of these agreements. Private oil companies were in no position to counter this situation with similar comprehensive agreements, which they were not authorized to make and which they did not have the power to fulfill.

In summary, the Soviet Union set before itself several objectives, both economic and political, in expanding its oil exports:

(1) Economic objectives
    (a) Rapid economic expansion
    (b) A continued rise in the standard of living and a resulting heavy requirement for consumer goods
(2) Political objectives
    (a) Economic and political penetration of weak, underdeveloped, and politically uncommitted countries
    (b) Increased dependence on Soviet oil trade by noncommunist economies, particularly those of the underdeveloped and semideveloped nations.[22]

Soviet Oil Exports to Noncommunist Countries

From its beginning, Soviet foreign oil trade with the noncommunist nations had been channeled in two major directions: one to Western Europe and all other industrial nations, and the other toward the developing nations of Asia, Africa, and Latin America. Being a highly merchantable commodity, Soviet oil was used in Western Europe in order to obtain much-needed equipment. In the underdeveloped parts of the world, oil trade was unabashedly used as a tool to increase Soviet political leverage upon those nations.

In competing for markets both in Europe and in Asia and in Africa, the Soviet Union had been competing with the Western oil companies. In communist doctrine, these oil companies were a symbol of the capitalist system. They now became the economic rival and the target of Soviet oil ventures abroad. In the Soviet Union, these oil companies

were represented as the epitome of the "entire edifice of
Western political influence . . . of all military bases and
aggressive Blocs." The destruction of these large monopolistic enterprises was a precondition to the collapse of
Western economy. "If this foundation cracks, the entire
edifice may begin to totter and then come tumbling down."
This view gave Soviet oil trade an added dimension inasmuch
as it was not merely intended for economic gain or for positive increment of political and strategic advantages but
also "to disrupt, undermine and, if possible, destroy the
position of the private oil industry."[23]

Soviet Oil Exports to Western Europe. The bulk of Soviet oil had always been shipped, via pipelines or tanker,
to Western Europe. In 1962 and 1963, it amounted to two-
thirds of total Soviet foreign oil trade. In 1962 noncommunist nations purchased $476 million worth of oil from the
Soviet bloc, thereby representing 9.2 percent of all their
imports from bloc countries; it also constituted the largest single import item from the area. Nearly 80 percent of
all this oil import went to six customers: Italy, Western
Germany, Japan, Sweden, Egypt, and Finland (see Table 16).


TABLE 16

Soviet Oil Sales to Selected Western
European Countries, 1961
(thousands of tons)

| Country | Crude | Petro Products |
| --- | --- | --- |
| Italy | 5,565 | 765 |
| West Germany | 1,620 | 2,505 |
| Sweden | 0 | 2,565 |
| France | 110 | 1,040 |
| Austria | 165 | 505 |
| Belgium | 0 | 340 |
| Denmark | 0 | 220 |
| United Kingdom | 0 | 110 |

Source: U.S. Congress, House of Representatives,
Committee on Foreign Affairs, Soviet Economic Offensive in
Western Europe, 88th Cong., 1st sess. (1963), p. 13.


More than 50 percent of the total sales was regularly
purchased by NATO member countries. The largest corporate
customers were government-owned companies, 50 percent of

which were unintegrated enterprises. Among them, power
plants constituted 11 percent of total sales.

Soviet Oil Exports to Less Developed Nations. The
major objective of Soviet trade with the less developed
areas was to exert political influence by economic means.
To this end, the Soviet Union extended to these nations
most-favored trade conditions that were not within the
reach of the competing private companies. Under these favorable conditions, Soviet trade with the less developed
nations soon surpassed that of the industrialized nations.
The following table shows the 1961 trade figures as a percentage of the 1955 figures.

| | Industrialized Nations | Communist Nations |
|---|---|---|
| Import by the less developed nations | 80 | 121 |
| Export by the less developed nations | 148 | 81 |

Thus, while the industrialized nations were absorbing
more and selling less in their trade with the less developed areas, the trade with communist bloc nations indicated
an opposite trend in which imports exceeded exports. In
terms of total Soviet foreign trade, the trade with the
less developed nations figured in 1961 as 28 percent of
total bloc trade. Out of this, the Soviet Union purchased
mainly agricultural commodities and raw materials, in exchange for which it exported to these countries machinery,
petroleum, food, and ferrous metals. None of the commodities purchased from these areas was as important to the Soviet economy as the products exported from the industrialized countries. This kind of trade was untenable on purely
economic grounds. Obviously, other considerations than commercial ones motivated it. Evidently, the Soviet Union expected substantial political compensations in return.

Soviet aid to petroleum activities in the less developed areas of the world concentrated mainly on fostering
the development of state-owned enterprises that could compete directly with the activities of local and international
free enterprise organizations. Soviet aid to such endeavors
was extended in the form of long-term loans at very low interest rates. Long-term credits for as many as 12 or 15
years were quite common. The favorable terms under which
these clients were allowed to borrow included (1) repayment
by installment after completion of project so that installment payments could be made out of profits already made;

(2) interest rates as low as 2.5 percent, thus undercutting
Western bank loans at 5 and 6 percent by more than one-half;
and (3) in the absence of hard currency, repayment to the
Soviet Union made either in the form of local commodities
or even in local "soft" currency. The Soviet Union especially favored the encouragement of construction of state-
owned refineries. In this way, while purchasing Soviet petroleum machinery and equipment for their new refineries,
these countries could not only manufacture their own petroleum products in competition with the Western companies but
they also could order their crude oil directly from the
Soviet Union.

The largest part of Soviet assistance to less developed nations in support of state-owned enterprises was concentrated in countries embracing the Middle East and the
Indian Ocean. Primarily concerned with protecting its own
southern flanks, the Soviet Union offered assistance to
Iran and Afghanistan. Nevertheless, its support of projects in Ceylon, Indonesia, and India could be ambiguously
construed to mean either a move in the overall pattern of
Soviet economic offensive or a strategic maneuver to con-
strain and divert the West further away from the southern Soviet border. It might be concluded that while Soviet
economists were thinking in terms of a commercial expansion,
Soviet strategists were advocating a similar pattern for
reasons of national security.

Thus, in 1956, the Soviet Union made a gift to the government of Iran of its own portion in the holdings of the
Soviet-Iranian Stock Company, a corporation established for
the purpose of prospecting for oil in northern Iran, on condition that this concession would not be reassigned to any
other country or foreign organization. This move accomplished two things for the Soviet Union: It created the im-
pression of good will as part of the general pattern of relaxation of international tensions generated at that time
by the new regime at the Kremlin; it also safeguarded Soviet
strategic requirements against the presence of foreign Western powers in the proximity of its vulnerable southern border.

The Soviet gift proved to be too big for the Iranian
government to swallow, since the Iranian treasury was in no
position to begin geological explorations and drillings in
northern Iran, and since all non-Iranian investors were excluded from the project, there was no one else in a position
to put up the necessary initial capital. As a result, no
progress was made in the northern Iranian oil prospecting.
Again, in 1959, the Soviet Union assisted Iran in offering

to prospect for oil in northern Iran and to give its government 15 percent of any petroleum found there if, in exchange,
the Teheran government would promise not to permit any foreign military bases in its territory.  This offer was shrewdly devised both in relation to the commercial aspect of the
proposed deal as well as for strategic considerations.  Commercially, Western concessionaries at that time were already
beginning to pay local governments as much as 50 percent in
royalties out of net earnings.  Taking advantage of the
terms of the agreement that prohibited foreign concessionaries from exploiting the mineral resources of northern Iran,
the Soviet Union reappeared again not as a concessionary
agency--a term to which these oil-producing countries attached an "imperialistic" stigma--but as a mere outside
"contractor" employed by the government of Iran.  The commercial terms of the deal took advantage of the fact that
the Iranian government could not employ any other "contractor" for the purposes of oil exploitation in northern Iran.
Strategically, by this deal the Soviet Union not only would
have been assured that no Western interests would be present in northern Iran but also that Soviet interests would
be replacing them independently in this territory vital to
its national security and without risking the traditional
stigma of imperialism commonly attached by left-wing parties to Western petroleum enterprises abroad.

In the same year (1959), the Soviet Union also extended a $140 million credit to Iraq for the construction
of various projects including a chemical fertilizer plant
with a 60,000-ton annual capacity and a geological survey
for promising petroleum prospects in the entire country.
Soon, shipments of geological equipment began to arrive in
Iraq, and a large repair shop and laboratories were established by the Soviet Union to service this equipment.  In
addition, the Soviets financed the construction of a new
refinery of 40,000 barrels a day (b/d) capacity near Basra
on the Persian Gulf.  These subsidies were climaxed when,
in 1961, Iraq suddenly cancelled all unused concessions belonging to Western companies and instead placed its own order with the Soviet Union for a complete rig at a reported
value of $1.68 million.[24]  This move indicated the extent
to which the Soviet Union was willing to go in support of
state-owned oil enterprises in the less developed areas in
order to displace the Western concessionaries.

Similar support was extended to the Syrian government
for the construction of a refinery with a 20,000 barrels a
day capacity at Homs, which was followed by an agreement
to undertake the project of oil prospecting in this country.

As a result of these efforts, oil was discovered in 1959 at Qaram Chok and, in 1963, at El Ejesire.[25] Another project for oil prospecting in the Middle East was undertaken by the Soviet government in Yemen.

In Africa, Soviet support of state-owned oil enterprises was chiefly valuable in Egypt, Ethiopia, Ghana, and Mali. In January 1958, the Soviet government extended a $175 million trade credit to Egypt out of which the petroleum industry received the major part of the loan. This portion was invested in geological surveys, technical assistance, and equipment. It also paid for the construction of a lubricating oil plant at Suez with a 1,200 b/d capacity; a refinery, also at Suez, with a 20,000 b/d capacity as well as a desulfurization plant with a 4,800 b/d capacity; a gasoline plant with a 6,000 b/d capacity; and a plant for production and repair of prospecting and drilling equipment used in the Red Sea, along the Mediterranean coast, and in the Gulf of Suez.[26]

Soviet support of the Ethiopian petroleum industry began in 1959 with a $44 million loan designated for geological surveys, appropriate equipment, and technical assistance needed for the construction of a 10,000-b/d refinery at Assab on the Red Sea, at a cost of $12.7 million. In March 1960, a team of Russians began extensive geological surveys.

While major Soviet efforts in this continent were concentrated in East Africa--in Egypt and Ethiopia, both of which had outlets to the Red Sea--some minor Soviet loans for oil enterprises were also extended to nations in West Africa. Ghana received from the Soviet Union a $40 million long-term credit for oil exploration and a supply of petroleum products, in exchange for which Ghana was expected to make repayments of the loan in the form of its own traditional commodity exports. At the same time, 25 Soviet oil geologists arrived in March 1962 to begin geological surveys.

Prospecting for oil, gold, and diamonds under a Soviet assistance program had also begun in Mali in 1961. This project was supplemented by provisions to ship large quantities of Soviet petroleum products to Mali. Total Soviet loans to Mali under these provisions amounted to $44 million and extended over a period of 12 years.[27]

Traced on the map, however, it was seen that the bulk of Soviet aid to less developed nations ran in one consistent and uninterrupted direction. In Africa, most of the Soviet aid was granted nations located along the Red Sea or those situated in a position to control its approaches. In Asia, Soviet support of state-owned petroleum projects extended in one uninterrupted crescent from Iraq and Iran on its immediate southern border, through Afghanistan and Pakistan, to India and Indonesia in the Indian Ocean.

The loans to Afghanistan were the earliest Soviet ones extended to less developed nations after Stalin's death. It had also been the first country to receive Soviet economic aid in the postwar era. In 1954 Afghanistan received a $1.2 million Soviet loan for the construction of three storage tanks, while in 1955 the Soviet Union granted Afghanistan $2 million for the purchase of asphalt to pave the streets of Kabul. Between 1958 and 1960, the Soviet Union added nine storage tanks and, in January 1962, after natural gas had been discovered in this country, commenced geological surveys and prospecting and constructed plants for petroleum and chemical industries, culminating its effort in October 1963 with the construction of a gas pipeline from Afghanistan to the Soviet border, capable of supplying the Soviet Union with 1,500 million cubic feet of gas per year.[28]

Soviet contributions in support of the petroleum industry of Pakistan during the Khrushchev era were limited to a $35 million credit loan, extended in January 1962 for a period of 17 years at a low rate of interest. According to the provisions of the agreement, the Pakistani government was to retain rights to all minerals discovered under this plan. One year later, in January 1963, a team of 79 Soviet technicians arrived in Pakistan as provided by the agreement.[29]

Completing the crescent of commercial loans and strategic grants that began to dot the political map of Asia, the Soviet Union extended its technical assistance and a $35 million credit to Indonesia for the construction of a refinery operating at a capacity of 30,000 b/d in Java and for the exploration of oil fields on that island.[30]

It is apparent from the above that the nations rewarded by Soviet aid were selected neither at random nor according to their need. The pattern of these grants and loans suggests that the beneficiary nations were singled out because of their geographical location. Except for short-lived assistance to state-owned petroleum enterprises in Ghana and Mali, which coincided with the Congo incident, all beneficiary nations were geographically located along the direct road to the Indian Ocean, whether on the overland route via Afghanistan and Pakistan or along the Red Sea coasts.

Commercial reasons did not appear to have motivated such a reallocation of economic benefits. Not only the

180

selection of the recipients but also their treatment seemed
to indicate a noneconomic motive. For theoretically, at
least, by suddenly abrogating contracts or interrupting
supplies unilaterally and arbitrarily, the Soviet Union was
able, if it so desired, to exert a great measure of political influence upon those nations that had become dependent
on this trade.

## Soviet Oil Exports to Communist Bloc Nations

While rebuilding its oil industry in the early postwar
years, the Soviet Union, being temporarily dependent on
foreign sources, confined its fuel imports mainly to Ruman-
ian oil and Polish and East German coal. Energy supplies
were thus sustained within the Soviet orbit, independent
of Western influences on international market fluctuations.
When Khrushchev reoriented Cold War policies toward
the path of economic competition with capitalistic nations,
petroleum became his main weapon. Petroleum made possible
Soviet incursions into established European oil markets as
well as the less developed areas of the world. To accomplish these, the Soviet Union undercut Western standard
prices, offered attractive barter agreements, and accepted
local commodities and even "soft" currencies. Such a policy was very expensive. In order to be able to afford it,
the Soviet Union was impelled to increase the prices
charged for oil sold to communist bloc nations, achieving
in the process a viable overall balance sheet for its oil
industry. By 1962, the price average charged in bloc countries per barrel was more than double the price the Soviet
Union charged its noncommunist customers (see Table 17).
During the Khrushchev era, the East European nations
paid from 77 percent in 1957 to 219.2 percent in 1962 over
the price asked from noncommunist nations. This steep
price increase was due to the fact that as a group the East
European nations shifted from a net exporter of petroleum
to a net importer in 1959, and by 1962 they had become a
substantial importer. Finland and Yugoslavia paid intermediate prices between those charged of the two blocs.
However, these figures become more meaningful when consideration is given to the fact that distance from the source
of supply, which increases transportation cost of exports,
is always attributed, in free market calculations, to the
price differential of petroleum. In its intent to capture
the Western markets, the Soviet Union overlooked the basic
free market profit margin imperative and used its centralized trade control to force the satellites to pay for its
policies.

TABLE 17

Average FOB Export Prices for USSR Crude Oil
(dollars per barrel)

| Year | To Free World | To Satellites |
|------|---------------|---------------|
| 1955 | 2.16 | 3.38 |
| 1956 | 2.17 | 3.30 |
| 1957 | 2.55 | 3.28 |
| 1958 | 2.08 | 2.97 |
| 1959 | 1.68 | 3.01 |
| 1960 | 1.56 | 3.01 |
| 1961 | 1.38 | 2.97 |
| 1962 | 1.36 | 2.98 |

Source: National Petroleum Council, Committee on the Impact of Oil Exports from the Soviet Bloc, Impact of Oil Exports from the Soviet Bloc (Washington, D.C., 1962), p. 7.


SUMMARY AND CONCLUSIONS

U.S. Kremlinologists, viewing the Soviet scene from their armchairs, were greatly impressed by the message of détente that was said to have come out of the 20th Congress of the Communist Party of the Soviet Union. Khrushchev's message to the world on "peaceful coexistence with economic competition" was picked up by every Sovietologist in this country, but only half of it was studied, explored, and speculated on. The first part of the Soviet slogan absorbed the attention of all those scholars, while very few of them advanced their research to the second part in order to research its practical and theoretical implications. Thus, the slogan that was important exactly because of its second part, which pointed out the path of future Soviet foreign policy in regard to the imperialistic nations, was, in fact, studied almost exclusively in reference to its first part, which was a mere reiteration of well-known Leninist and Stalinist dogma. Both Lenin and Stalin had recognized on various occasions that objective conditions in our particular phase of materialistic history dictated a pause in the struggle toward a world revolution and a state of détente with the capitalistic nations. This con-

dition, among other names, was also called "peaceful coexistence."   To Lenin this historical interpretation of the
revolutionary path justified the New Economic Policy (NEP);
to Stalin it was the foundation on which he constructed his
purely nationalistic policy of building socialism in one
country.   Since good communist form required everyone to
precede an original thought by reference to the giants of
the past (Marx, Lenin, or Stalin), it is not surprising
that Khrushchev's announcement of his new policy of "economic competition" with the West was prefaced by a routine
quotation from those scriptures.   It is most unfortunate
that U.S. Kremlinologists never progressed from the first
part of the slogan to its second part.

In justification of this scientific omission one ought,
however, to keep in mind the many historical events that
seemed to increase the significance of this statement by
their contradiction from the avowed policy.   In the prevailing Cold War conditions and mounting international tensions
that had existed in relations between the two blocs since
the end of World War II, the Soviet view of the division of
the world into two camps was only enhanced, in U.S. minds,
by the Berlin crisis and the Korean War.   In addition, many
Americans also suspected hidden motives behind Soviet manip-
ulation of foreign communist parties and front organizations
and Western peace movements and Soviet exploitation of such
trouble areas as Iran and Guatemala.   In these conditions, a
message from the Kremlin that referred to "peace" was generally welcomed by everyone.   The subsequent revelation of
Khrushchev's de-Stalinization speech further confirmed many
U.S. scholars in their opinion that the Soviet Union was,
indeed, on the threshhold of a major departure from past
policies.

Future events, however, seemed to conflict with this
wishful analysis.   In the year of "peaceful coexistence"
and de-Stalinization, the Soviet Union also invaded Hungary
and effectively intervened in the Suez Canal crisis.   In
the following year, it sent up Sputnik and thus led the way
to the "missile gap" scare in the United States.   This scare
was not completely over, and the second Berlin crisis was
still painfully fresh in mind, when Khrushchev created more
international tension with the Cuba missile crisis.   Thus,
U.S. Kremlinologists were so busy trying to reconcile the
message of détente heralded by "peaceful coexistence" with
actual Soviet policies that they had no time left to look
into that part of the same statement that also announced
"economic competition" as a method of settling the unrecon-

cilable contradictions between the two camps and winning a
victory by means other than war.

As on many occasions in the past, the Soviet Union
again looked to the developing nations of Africa and Asia.
"The policy of peaceful coexistence," says the Statement of
the 81 Communist and Workers Parties, "is a policy of mobil-
izing the masses and launching vigorous actions against ene-
mies of peace. . . . In conditions of peaceful coexistence
favorable opportunities are provided for the development of
the class movement of the people of the colonial and dependent countries." Thus, a new concept was born in the communist vocabulary: neocolonialism. This concept was applied to U.S. dollar imperialism in the developing parts of
the world but never to Soviet trade and aid policies in
these countries.

U.S. preoccupation with the doctrinal interpretation
of "peaceful coexistence" resulted in a large number of
books and research papers on the subject, many bordering on
sheer sophistry. On the other hand, scholarly works on Soviet economic competition were few in number and more often
than not were written with an economic rather than a political bias. If research into Soviet foreign trade and aid
policies was limited both in scope and in quality, the research undertaken by scholars into Soviet foreign oil policy was nearly nil. The material that does exist on the
subject comes primarily from concerned commercial quarters
as well as official reports made by hired investigators to
provide cursory background information to executive agencies and legislative committees. Again, these reports were
undertaken by men trained primarily in economics and having
little if any knowledge of political concepts in general
and international relations in particular.

However, with the repeal of many of Khrushchev's policies and the withdrawal of the Soviet foreign oil trade offensive, the impact of this episode of history upon the
world economy disappeared. But its lesson should remain
imprinted, as a warning, upon U.S. foreign policy, not for
what it is now or even for what it was in 1956-64, but for
what it can become as a weapon of the Soviet leaders at
any time they decide once again to reverse their policies.
At a later date the use of oil as a political weapon was
urged upon the Arab nations in 1973. At that time Radio
Moscow's broadcasts in Arabic to the Arab world castigated
Arab leaders for not using oil as a political weapon against
"Western imperialism" in the first place and against "Zion-
ist aggression" in the second place. The ordering of priorities by Radio Moscow clearly indicated its objective in
hurting Western interests.

By the end of the 1950s, Soviet foreign trade in petroleum had become the most significant economic factor influencing contemporary politics. No other single commodity
could rival the role of Soviet oil in the world market and
its effects on new political alignments. The rapid momentum of its growth at the expense of established enterprises
in traditionally Western markets caused a great deal of
anxiety in many quarters for economic as well as political
reasons. Most concerned, economically, were the international petroleum companies, the refining industry, and, of
course, the petroleum-exporting countries of the Middle
East and Venezuela, whose national economies entirely depended on the revenues they derived from oil and its products. On the other hand, the United States, not being completely oblivious to the economic effects on nations whose
friendship it was seeking, was, in effect, chiefly concerned with two other aspects of the same problem: Strategically, Western Europe's new, though limited, dependence
on Soviet oil threatened, to the extent of the dependence,
to weaken NATO's measure of preparedness; and, politically,
the Soviet petroleum trade also threatened to create a new
set of alignments in Asia and Africa. As a result, two
concurrent movements came into existence simultaneously:
While the U.S. government was pressuring the NATO nations
to enforce controls on their petroleum imports from the
Soviet Union, the oil-producing nations of the Middle East
and Venezuela were organizing a mutual policy-making association whose task was to promote their own trade in oil
and oil products vis-à-vis that of the Soviet Union. The
two movements, although complementing each other in their
effort to curb the Soviet petroleum offensive in the noncommunist world, were not actually working toward the same

goal. While the Middle East nations were primarily concerned with posted prices and royalties, but not shying at
the same time from using their economic leverage for political objectives, the U.S. government worried mainly about
Western security and the political and strategic needs to
free it from excessive dependence on any hostile sources of
supply, either communist or Arab. The understandable Arab
obsession with revenue from petroleum production and the
emphasis of U.S. policy on diversification of petroleum resources toward friendlier areas resulted in a sharp conflict of interests. Caught up in the midst of this new
complex of international tensions were the petroleum companies, which were seen as Western agents by their Arab
hosts.

This chapter deals with five reactions to Soviet intervention in the international oil market. The chapter's
first section depicts the problems encountered by the oil
industry due to Soviet competition. It also cites responses
by some leading oil industrialists to the Soviet threat and
its relation to the national interest of the United States.
Finally, the section traces the early notions of opening up
channels for East-West trade as they were expounded at that
time by the private sector. The second section views U.S.
political and military interests in the Mediterranean and
the Middle East as distinguished from private U.S. oil concerns in that area. It also describes efforts made by the
U.S. government to influence international oil trends by
enacting new legislative measures at home and through discussions with the European Common Market Council and the
NATO allies abroad. The third section deals with measures
taken by the oil-producing nations to counter the recessive
trends in the world oil market that were set in motion by
Soviet competition.

The fourth section describes how Europe, once again
spurred by oil crises not of its own making, enlisted its
own industrial and commercial know-how in order to develop
further its important ancillary industry of refineries and
petrochemical plants, and how it has attempted (unsuccessfully thus far) to formulate a European oil policy that
would increase independence from foreign oil concerns primarily by diversification of suppliers (a move that encouraged the shift toward larger national oil companies in many
of these countries) and by creating a platform for a petroleum common market. The fifth and last section deals with
the ever important ancillary tanker industry and its bearing upon ocean transportation on the eve of the six-day war
(1967), which interrupted all water-borne traffic between
the east and west across the Suez Canal.

# THE INTERNATIONAL COMPANIES' DILEMMA

To U.S. eyes, anxiously following the rapid growth of Soviet petroleum trade, it seemed that the need claimed by the Soviet Union for foreign exchange did not represent the whole truth, in view of the fact that the Soviet Union had gold reserves believed by some observers to be as much as $8 billion.[1] Instead, many prominent leaders of U.S. industry interpreted the commercial expansion of Soviet petroleum as being a tool in the hands of the Kremlin to disrupt Western trade, interfere with its sources of raw material supplies, and win the struggle between the two blocs by economic means.[2]

In 1960, a U.S. "oil delegation," composed of businessmen in the petroleum industry, visited the Soviet Union. These experts were impressed by the size of the pipelines that were being prepared at that time in Russia to transport substantial shipments of crude oil easily and quickly into noncommunist markets.[3] Intensive developments in the Soviet petroleum industry indicated that further expansion was anticipated. New oil fields were being explored, a pipeline network was being constructed, and an expanded tanker fleet was being assembled. Based on all these developments, U.S. petroleum forecasters during that period estimated that Soviet export capacity would rise from 600,000 b/d in 1965 to 1.2 million b/d in 1970, replacing in the process an increasing portion of other petroleum sales to noncommunist nations.* Even then, the volume itself of these exports from the Soviet bloc could not have had a major significance on the international market, as the demand by noncommunist nations for petroleum supplies was expected to continue to rise at a much faster pace than Soviet foreign trade. The following figures show estimated petroleum demand (in millions of b/d) by noncommunist nations for selected years.[4]

| Year | Amount |
|------|--------|
| 1938 | 4.7 |
| 1955 | 14.4 |
| 1960 | 18.8 |
| 1965 | 24.4 |

---

*This prediction did not materialize, because of the reversal of Soviet foreign trade policy, particularly in relation to oil, as described in Chapter 8.

It was also generally expected that by 1965 the petroleum demand of the consuming nations outside the Soviet
bloc would exceed that of the United States and thereby
would become the largest existing market in the world.  It
was thus imperative for all producing nations as well as
trading companies to protect the availability and the stability of this future market.  According to remarks made by
Secretary of the Interior Stewart Udall in 1961, even if
Soviet petroleum trade did not constitute a present or even
a potential threat in volume, what was a major concern was
the Soviet ability "to exercise a degree of price flexibility" that had "an effect disproportionate to" the quantity
of their oil.[5]  And, according to a U.S. producer,

> The operations of a competitive economic
> system mean that even 3% or 4% of low
> priced crude oil causes the entire price
> level to decline.  When Russian imports
> reach 9% as they do in Germany, 15% in
> Sweden and 21% in Italy, the entire price
> level tumbles to the Russian level.[6]

U.S. businessmen having investments in the Middle East
and attempting to enlist public opinion support in the United States, attacked Soviet oil policy from every possible
angle:  patriotic, strategic, economic, and moralistic.
Appealing to U.S. patriotism, M. J. Rathbone, president of
Standard Oil Company of New Jersey, expounded on the great
significance of the developing nations to U.S. security and
welfare, thus tying together private commercial interests
and the nation's safety:

> We know that America's defense no longer
> lies within our own borders but now de-
> pends on friendly neighbors and the main-
> tenance of a military system capable of
> acting quickly and forcefully in any part
> of the globe.  To back up this conviction
> we have built military bases abroad, have
> given billions of dollars to strengthen
> the economies of our friends and allies
> and have paid the dearest price of all by
> spending American blood in such far away
> places as Korea.  We know now that what
> happens in other countries is definitely
> of great military and economic importance
> to us.[7]

Similarly, the Committee for Economic Development (CED), a private organization composed of presidents of universities and presidents and chairmen of the boards of the largest corporations, alerted U.S. public opinion to the immense importance of the developing areas for the vital interests of the nation:

> The first [reason], of course, is our military security. It is they who have made possible the bases giving us the deterrent striking power which Russia cannot duplicate vis-a-vis the United States. . . . That is the "why" in the military sense. . . .
>
> [On the] economic side of it . . . the United States is no longer self-contained in raw materials. . . . We are heavy importers of petroleum, iron ore, bauxite, and paper, and it is important to us in the future to assure these supplies.
>
> The great undiscovered raw-material assets of the world lie in the underdeveloped countries. [It would be] a matter of ordinary prudence that when those discoveries were made they should be available to future American industry and not be cut off by the lowering of a curtain.
>
> And then the third point, of course, is that there lie the great markets of the future for American goods.[8]

Morally, the Soviet petroleum offensive was interpreted by spokesmen of the private oil companies as a reflection on the fact "that state trading in oil had all the vices which were alleged to be a part of cartel organizations, compounded by the use of political power."[9] Not only that, but it reflected also the "ironic situation that a socialist nation in competing with the West resorted to the most abusive of cartel practices--'Keep the prices down to drive out competition and when you get the market control--raise prices.'"[10]

In March 1963, George F. Getty, II, president of Tidewater Oil Company, summed up the threat of Soviet aggressive petroleum policy as follows:*

---

*The "threat" on the domestic scene refers to the then proposed reduction of percentage depletion tax provisions to U.S. oil producers.

> The major problems facing the petroleum in-
> dustry today are not technical but political
> and the most ominous in the industry's his-
> tory.  The two major political threats are
> the cutthroat competition of the government
> controlled Soviet oil industry, and the
> equally formidable menace of government con-
> trol on the domestic scene.[11]

There was, however, one more national problem, which
Getty failed to mention and which was clearly the making of
the oil industry itself.  It was the problem of the balance
of payments, which, at that time, had reached a threatening
point where it could imperil the entire economy of the country.  Early in 1965 President Lyndon B. Johnson initiated a
Voluntary Balance of Payments Program, according to which
U.S. corporations were urged "to do what they can to help
stem the flow of gold abroad by slowing down the pace of
overseas expenditures, by returning profits and other funds
to this country promptly, and by raising more investment
funds abroad."[12]

The international petroleum industry's immense investments and large expenditures abroad immediately brought the
industry to the attention of those government officials who
were seeking ways and means to adjust balance-of-payments
deficits and who asked the companies to restrict their
1965-66 investments to 90 percent of their expenditures
during 1962-64.[13]  The foreign oil corporations, however,
did not respond to the president's program.  To comply with
such a program, their spokesmen argued, at a time when the
world markets were vastly increasing and the competition
for them was mounting, would mean, in these companies' opinion, that they would have to deny themselves the funds for
construction and expansion and, thereby, would be allowing
these new markets to turn to oil companies of other nations
for their requirements.  They pointed to the fact that adequate supplies of foreign petroleum were important not only
to satisfy overseas markets but also in order to respond to
high-priority national strategic needs.  A glaring example
of this situation concerned certain military officials who,
seeking to contribute their share to help the balance of
payments problem, attempted to reduce their oil purchases
abroad but found it difficult to locate enough jet fuel in
the United States to meet their requirements.[14]  The special
predicament of the international petroleum industry was
summed up in the September 1961 issue of World Petroleum as
follows:

The oil companies themselves are caught on
the horns of a dilemma:  they cannot lower
prices to meet Soviet price competition
without incurring the wrath of the producer
governments which look with disfavor on any
price cutting; they cannot combine nor act
in concert to meet the threat due to antitrust legislation, and that effectively
hamstrings their efforts.  Any action taken
would appear to have international implications that preclude the oil companies themselves from effectively meeting the threat.
This leaves it up to the importing countries themselves.  There is reason to suspect that they will act with caution, having full knowledge of the political strings
attached to Soviet oil and the need for reliability as complete as it can possibly
get in a continuing dependable oil supply.[15]

Unable to operate on a political level themselves,
the international petroleum companies called for cooperation among the governments of the consuming nations, especially those who were members of the NATO alliance and
therefore had a clear vested interest in keeping their
economies independent of Soviet influences and possible
future blackmail.

The remedy is to put a country by country
percentage limitation on Russian imports
the same way the U.S.A. did on all crude
and product imports (roughly 15% of the
total demand). . . .
This problem of control could be
handled vertically, industry by industry,
or horizontally by the governments involved.  The government method is the better as it would create an apparatus to
handle any dumping or singling out of a
product or country as the need arises.
This is largely an alliance problem.
Just as the NATO military alliance has prevented Russia militarily from doing what
Hitler did in taking on one country at a
time--an economic alliance would perform
a similar mission--bloodlessly.

The West should use its present enormous economic and banking power to keep
these Russian forays to a reasonable price
level.[16]

At the same time when the companies were asking the
government's active intervention in influencing the NATO
allies to curb their oil purchases from the Soviet Union
for political as well as for military reasons, these same
companies also desired now to reopen East-West trade in the
opposite direction.  If the Soviet Union could sell in
Western markets, they reasoned, why could not they sell in
Eastern bloc markets?  Objectively, the communist bloc was
a good potential client.  As seen in Chapter 8, cut-rate
Soviet prices of oil offered in the West were being balanced by exorbitant rates charged friendly fraternal nations, which far exceeded those posted by the international
companies.  Due to Soviet political, rather than commercial,
calculations in their oil offensive on new international
markets, the Soviet bloc was left wide open to rival competition by companies whose markets the Soviet Union was now
invading.
CED, in discussion of East-West trade together with
French, German, Italian, and Japanese groups, supported a
proposal for the formulation of a common trade policy among
the industrial countries, beginning with the private citizens levels and bringing it up, gradually, if and when warranted, to the intergovernmental level.  Based on these
discussions, CED submitted its recommendation for government consideration to the effect that

The United States should be prepared to
take steps to expand trade selectively,
with individual Eastern countries, where
that seems economically and politically
beneficial.  The benefits must be calculated realistically and selectively with
regard to particular types of trade with
particular Eastern nations.  The United
States should cooperate with other Western
countries in exchanging information and
developing policy with respect to East-
West trade.  As part of the process of expounding trade the United States should
obtain from Eastern countries certain pro-
tections and opportunities for Americans
engaged in such trade.[17]

CED based its recommendations on the premise that, according to the Western system, "a private party in one country should be able to sell in another country when it is his best source of supply."[18] Such a system today, however, is not freely operative between East and West, because the "nature of the political relations" between the two blocs "affected the basis of trade in ways not applying to trade within the West."[19] Therefore, CED, an organization representing to a large degree the huge U.S. corporations under private ownership, condoned government intervention in this instance where political rather than commercial considerations seemed to be the prime motivating factors in the other bloc's calculations. "The economic systems of the communist countries," CED's report stated, "made the Western principle of 'minimizing direct government intervention' meaningless in reference to them."[20] The scope of government intervention in this commerce should be confined, then, to the negotiations of "agreements of limited duration with individual Eastern countries, as seems economically and politically beneficial in the light of both trade and non-trade advantages obtained by the United States."[21]

## THE LIMITATIONS ON GOVERNMENT POWERS

In Washington both the executive branch and the Congress, for political and military reasons, were gravely suspicious of the Soviet economic offensive, its penetration of international markets, and its challenges. It was once again remembered in this connection that Soviet ambitions in the Mediterranean and the Middle East dated back to the old Czarist regime. These ambitions had been reaffirmed in Molotov's communication of November 25, 1940, to the German Ambassador in Moscow in which he had sought to partake in a division of the world with Hitler. Again, in the immediate postwar years, the Soviets had prolonged their military occupation of Iran, had sought a trusteeship of the former Italian colonies in North Africa, had been also suspected of extending support for the civil war in Greece, and had made the "satisfactory" solution of the question of Trieste a condition to the conclusion of the Austrian peace treaty. It was also noted in Washington, with increasing anxiety, that if those early Soviet attempts to gain greater influence in Iran and the Mediterranean had been unsuccessful, the subsequent efforts to penetrate the Arab world of the Middle East were bringing far greater political and

military dividends.  At the same time, the West's standing,
vis-à-vis that of the Soviet Union, was rapidly deteriorat-
ing.  NATO's position in the Middle East became limited to
Aden and Cyprus, the latter being torn by local antagonisms.
Moreover, the recent developments of Soviet missile capability imperiled those bases.  According to William Hessler,

> Because the locations of the bases are known
> and can be pin-pointed with precision, they
> are exceedingly vulnerable. . . . Thus po-
> litical and technological developments alike
> have forced us to write off a fair part of
> our investment in forward positions . . .
> [causing] the deterioration of America's
> strategic position in the Mediterranean.[22]

Several remedies were being offered in order to bolster the sagging political foreign policy of the United
States in the Middle East by military power.  The most obvious cure-all, designed to replace vanishing territorial
bases and diminishing political prestige, was the Sixth
Fleet, whose mission it became to protect U.S. interests
in the Mediterranean, the Middle East, and along the Red
Sea coasts as far as the Indian Ocean.  The Sixth Fleet reaffirmed the importance of the sea as a highway in NATO's
strategy.  Structured toward combined forces operations, a
mixture of land, sea, and air forces, enhanced by a striking capability from the carriers, the Sixth Fleet came under a joint staff command.  With the aim of achieving a
wide range of military and diplomatic options, "the great
advantage of the balanced force concept" was also handy in
widening "the range of the choice it gave to political
leaders."[23]
In addition to its high state of readiness, punctuated
by a nuclear striking capability, the Sixth Fleet offered
an important political advantage over land bases, inasmuch
as it could be utilized as a means by which to remove from
the United States the stigma of "colonial power" within
which it had been bracketed by Arab nationalism and its
anti-Western hostility.  In order further to allay Arab
fears, this policy, stressing the noncolonial nature of
U.S. presence in the region, had to emphasize the separa-
tion, if not the incompatibility, between the political
and military policies originating in Washington, and the
purely commercial interests of the private U.S. investor.
Within this international framework, the Middle East
oil corporations, facing the threat of Soviet competition,

were treading on ambiguous grounds. On the one hand, the administration and Congress were exceedingly concerned about the Soviet presence in the Middle East, but they were alerted to it by other reasons than economic considerations. Consequently, on occasion, there were situations where a political or a military solution did not support, and sometimes even contradicted, the commercial aspect of the same problem. On the other hand, again for a whole range of different reasons, among them the antitrust laws of the United States and the desire to improve the political relations with the Arab nations, the U.S. government was sidetracking any proposition that suggested a tighter coordination of policies with the international petroleum industry. There was no legal or practical way through which the administration, the Congress, and oil companies could meet the Soviet menace on a common footing of mutually reciprocal interests. Thus, while the president of the United States was issuing proclamations restricting the volumes of petroleum imports into the United States,[24] and while Congress was enacting bills and conducting investigations as to the nature of the Soviet threat and how best to combat it, both branches of the government were, in effect, limited in the range of their choices inasmuch as they were not the producers of international petroleum themselves, only the regulators (and even that only in clearly specified circumstances) of its flow within the national boundaries of the United States. At the same time, the international petroleum companies, while being able to initiate production and marketing policies to rival Soviet trade, were circumvented on two accounts: They were forbidden to act jointly for fear of violating the antitrust laws of the United States, and they were unable to go beyond the private commercial relations with other governments, as they lacked the diplomatic power and the authority to do so.

Several bills were thus enacted in succession by Congress that reflected, directly or indirectly, the state of the international petroleum market. In 1957, as a result of the acute shortages in tanker transportation experienced during the Suez Canal crisis, Congress passed a bill limiting owners' freedom to transfer their ships to foreign flags.[25] Accordingly, it became unlawful for U.S. citizens to lease, bareboat charter, or time charter for a period of more than one year or to mortgage or agree to sell to any person not a citizen of the United States any vessel unless it was "of no value to the defense of the United States" or unless the secretary of Commerce would testify in a public hearing as to the reasons that would compel such a transfer.[26]

The purpose of this bill was to have a larger number of ships available to serve U.S. policy in time of emergency. The frequent protests by U.S. shipowners whose vessels were registered under foreign flags that their fleets would constantly be available to answer any call from the U.S. government to serve the national interest did not seem to impress too many legislators. "To put it mildly," said Senator Warren Magnuson, "I am exceedingly skeptical as to the promised availability of these transferred vessels." Again, in 1960, the Merchant Marine Act of 1936 was amended to enact legislation removing the 50 percent limitation on construction subsidies awarded for vessels constructed in U.S. shipyards.

> These subsidies, which are paid directly to the shipyard, are designed to achieve parity in domestic construction costs with those of similar vessels constructed in lowest cost foreign yards. Recent maritime calculations indicate that the differential between domestic and foreign costs have exceeded the 50-percent limitation presently contained in the Merchant Marine Act of 1936.
> As a result of this, the competitive position which the principle was designed to accomplish is no longer present.[27]

More directly affected by current Soviet policies was the Trade Expansion Act of 1962 amending the original Tariff Act of 1930 in specifically disallowing the president "from entering into a trade agreement with the Soviet Union to grant them most-favored nation treatment; all importations coming from the Soviet Union are subject to the rather high rates that have been in force since 1930."[28] These and a number of other bills passed from time to time by Congress could not possibly meet the challenge of Soviet petroleum trade or neutralize the concern of the trading companies about their ebbing financial earnings.

There emerged a frequently articulated demand in many concerned circles for the regulation of coordinated controls upon Soviet oil imports by the European members of the NATO alliance. To this end, the U.S. government, through diplomatic channels of the Department of State and in meetings of the NATO command, began to exert its pressure on the allies for a common oil policy.

To U.S. eyes it seemed that the reluctance of the Western European nations to enforce tight controls over East-West trade was due chiefly to their inability to understand

the potential risks involved.  U.S. legislators concluded
that the NATO nations of Western Europe were unable to
achieve a "clarity of purpose" because they lacked the in-
sight into the basic issues as evidenced by the fact that
they chose to ignore Khrushchev's warning that under the
new policy of peaceful coexistence capitalism would be destroyed "through an ideological and economic struggle" and,
instead, these nations seemed to view Soviet penetration of
world markets as a step toward the "normalization" of trade
relations between East and West.[29]

The congressional special mission to Europe to investigate the effects of the Soviet economic offensive upon the
industrialized nations recommended that those nations should
coordinate a mutual policy of strategic-economic defense alliance as a counterpart of the mutual military defense alliance existing under the NATO agreement and in accordance
with the evolution of Cold War frame of references.  The
absence of such a policy, the committee maintained, could
very well play into Soviet hands inasmuch as it might con-
done the strengthening of the bloc's industrial strength
and military potential by means of uncontrolled practices
with the West, thereby increasing the bloc's capabilities
to spread further its political-economic penetration of underdeveloped nations.  Therefore, the NATO allies were advised to enlarge their list of controlled goods and to enforce it uniformly throughout the industrialized parts of
the world.

Several discussions of the problem of a common policy
took place at NATO headquarters.  As a result of these
meetings, in November 1962, the NATO command recommended to
its members that they should not sell large-diameter pipes
to the Soviet bloc.  This step did, in fact, "delay the completion of the Comecon pipeline and prolonged resort to rail
transportation," but, in itself, this measure was insufficient to control Soviet oil exports.  Further discussions at
NATO headquarters in regard to collective economic measures
were largely fruitless or ineffective.

Similarly, the Common Market commission discussed the
threat posed by the incursions of Soviet petroleum trade,
but the agreements reached at those meetings enjoyed only
a limited consensus pertaining, mainly, to controls of
sales to the Soviet Union of a narrow list of arms, atomic
energy material, and strategic materials.  No agreements
have been reached in regard to any type of technology and
other merchandisable goods, which nevertheless could contribute to the military industrial potential of the communist bloc.  In comparison with Western Europe, the United
States had adopted much tighter controls, which, however,
could be sidestepped by special export licenses.[30]

# THE ORGANIZATION OF PETROLEUM EXPORTING COUNTRIES

In spite of all U.S. attempts to curb trade of noncommunist nations with the Soviet Union, the industrialized nations of Western Europe and Japan continued to do business with the communist bloc. In return for oil and oil products, these nations shipped to the communists large orders of pipes, pumps, and compressors, which were used for the expansion and realization of Soviet petroleum goals.

Most affected by this incursion into the international petroleum markets by the Soviet Union were the Middle East oil nations and Venezuela. These nations had developed, within a very short time, from a secondary source of oil supply to the world's largest. Within a matter of 20 years the center of gravity of world oil reserves had moved from the Western Hemisphere to the Middle East (see Table 18).

TABLE 18

Noncommunist Crude Oil Reserves
(percent)

|  | 1938 | 1957 |
|---|---|---|
| United States | 59 | 17 |
| Middle East | 17 | 71 |

Source: Leonard M. Fanning, "The Shift of Petroleum Power away from the United States," a study report (Pittsburgh, Pa.: mimeographed, Gulf Oil Corporation, April 10, 1958), p. 2.

Since the 1950s, then, the principal suppliers of petroleum in the world have been Venezuela and the Middle East. Therefore, to the extent that Soviet bloc oil exports reduced the volume of exports of other exporting nations, Venezuela and the Middle East were the principal sufferers. Of these nations, Venezuela rated as the largest single foreign producer, with 29 percent of total free world foreign production. The four major Middle East producers were Saudi Arabia, Iran, Iraq, and Kuwait, whose combined production accounted for 44 percent of free world foreign total. In all these countries oil reserves were of crucial importance. Soviet competition could be destructive to their economies.

For reasons of their own, the Soviet leaders made no secret of their competitive designs. At the Second Arab Oil Conference, held in Beirut in October 1960, the chief Soviet delegate, Evgenii B. Gurov, director of Soyuzneft-export, the Soviet state agency charged with handling the sale of oil and oil products to foreign countries, made it clear that "Russia had every intention of exploring the European oil market in every way possible. As he put it, she was only seeking to restore her oil exports to the level they occupied before the second world war."[31] But as for the Arab countries, the Soviet delegate said that they did not need to fear regarding their own exports either to Western Europe or to other countries, in view of the fact that "world consumption of crude oil was continually increasing."[32]

"Russia's action may be legal," said Bustani, a well-known Lebanese businessman, "but undercutting a traditionally Arab oil market can hardly be regarded as a friendly gesture towards the Arabs."[33] It was his opinion, however, that the Soviet oil offensive was first and foremost a political maneuver to uproot Western interests from the Middle East.

> Russia does not need Arab oil for her own
> purposes, but it is clear that nothing could
> please her better than to possess the means
> of denying it to the West. . . . Hence her
> courting of the Arab world, her sometimes
> brilliant exploitation of the force of Arab
> nationalism, and her repetitive offers of
> aid and trade "without strings." . . . It
> is not, however, realized as widely as it
> should be that her own oil production may
> come to present an equal danger to the in-
> terests of the Arab oil producing and tran-
> sit countries. . . . It has, as a matter
> of fact, already led directly to reductions
> in the price of Arab crude oil.[34]

The U.S. oil corporations, faced with Soviet cut-rate prices, which overflowed the international market, were pressured into a price war with the Soviet Union. Without prior consultation with the oil-producing countries, the companies reduced the posted prices twice, in February 1959 and August 1960, averaging a reduction of $0.27 per barrel. Since, according to the oil-producing countries, these countries "were paid 50% of the net profits calculated on posted prices, this meant that a country with a production of 1,000,000 barrels per day lost . . . a total of $135,000 a day since the price cuts,"[35] with obviously bad effects on the Arab economies. It was this reduction of posted prices by the companies, rather than the Soviet threat to the whole market structure, that had finally united the oil-producing nations. Spurred by the urgings of Perez Alfonzo of Venezuela and Sheikh Abdullah Tariki of Saudi Arabia, delegates from Venezuela, Kuwait, Saudi Arabia, Iran, and Iraq met in Baghdad, where, on September 14, 1960, they created the Organization of Petroleum Exporting Countries (OPEC) as an "instrument for regular consultation and policy coordination among Member Countries."[36] Qatar, Libya, and Indonesia soon joined the organization.

The first task of OPEC was a domestic one--namely, to enact legislation decreeing, by mutual agreement among the member countries, some basic changes in the various features of the major concessions. These included the following:

1. The duration of the concessions was shortened to a period of between 25 and 45 years.

2. The royalty was now considered as an expense on the part of the company and not as a credit against income tax.

3. The new total payments, including royalties and income tax, were not far in excess of the 50:50 profit-sharing arrangement of the major concessions.[37]

OPEC's major foreign policy concern was to stabilize oil prices, principally by curbing the companies' power unilaterally to reduce posted prices. A second objective of foreign policy before OPEC was the dilemma of the removal of refineries from the oil-producing countries and their relocation in the oil-consuming countries.

Several roles were envisaged for OPEC at its inception, in addition to that of stabilizing posted prices. The enlargement of both property and power of the national oil companies was justified by OPEC on grounds of the Soviet example, where a state-owned agency enjoyed "competitive advantage in the international markets because it was backed by the State."[38]

The emergence of OPEC created several problems for the oil companies. In the first place, there was the question of the new concessionary setup. The old principles--legal, political, and even practical--could not be applied any more. They were replaced by new legal codes that developed in the individual producing countries independently of each other and replaced the uniform customs that had existed before. This breakup of concessionary practices, rooted in individual legal systems, was also reflected in the political atmosphere within each country, the spirit of nationali

and the anti-Westernism of the people. Thus, companies
were made to be wary of many codified as well as subtle
varieties of exploration prospects in foreign lands,[39] with
the result that an increasing number of them began to opt
for opportunities in friendlier areas where continuity of
contracts would be guaranteed and where they would not be
faced with sudden new laws affecting the industry or abrupt
and impulsive nationalization of all proprietorial rights.
In most cases, the decision in preference of other prospecting areas involved much larger investment capital, as
nowhere in the world were there oil lands that gave the
same high ratio returns per dollar as in the Middle East.
Nevertheless, weighed against the future security of their
new investments, companies as well as governments engaged
in petroleum explorations started to turn their sights toward other, even if more expensive, lands.

The trend in the international petroleum industry to
explore opportunities in friendlier and safer areas of production did not, in itself, alleviate the pressures exerted
by Soviet oil exports on the world market. The petroleum
industry was still looking for solutions. Since OPEC
seemed to be interested chiefly in the financial earnings
of the producing countries, and since the companies had no
political authority to act on a diplomatic level and since,
moreover, the U.S. government seemed to have only a limited
jurisdiction to step into an area that was assigned by law
to private enterprise, the U.S. industrialists began to believe now that controls on Soviet imports should originate
in the consuming countries themselves.

Gordon W. Reed, chairman of Texas Gulf Producing Co.,
regarded the situation largely as an alliance problem along
economic lines, parallel to the NATO military alliance.

> The problem of control could be handled ver-
> tically, industry by industry, or horizontal-
> ly by the governments involved. The govern-
> ment method is the better as it would create
> an apparatus to handle any dumping or sing-
> ling out of a product or country as the need
> arises.
> This is largely an alliance problem. . . .
> The West should use its present enormous
> economic and ranking power to keep these Rus-
> sian forays to a reasonable price level.[40]

In short, what Reed was asking for was an "economic
NATO" involving a plan of action for "an alliance of importing countries to impose quotas on Russian oil imports."[41]

In spite of the seemingly self-evident interest of the
Western European nations in minimizing their dependence on
Soviet sources of energy supply, the common good gave way
to the national interest of individual members of the OEEC,
who chose to profit by this type of East-West trade while,
at the same time, protecting themselves through vastly enlarged investments in oil explorations in the Middle East,
North and West Africa, along the Red Sea coast, and in the
oceans of the world.

WESTERN EUROPE'S RESPONSE TO THE
SOVIET CHALLENGE

The May 1956 Report of the Commission for Energy under
the chairmanship of Sir Harold Hartley (the so-called Hartley report),[42] which dealt with oil industry and consumption in the OEEC area during the postwar decade, studied
the difficulties of cost and security that might arise in
the future from Europe's growing dependence on imported
fuel.  The Suez Canal crisis, which occurred shortly after
the report was published, encouraged the prevalent policy
orientation toward the diversification of overseas sources
of supply and stock-building within the OEEC area as an insurance against a similar emergency in the future.  It was
in this light that the Robinson report,[43] which studied the
past effects and future implications of the Suez Canal petroleum crisis in the European economy, made its recommendations.
    Due to the 1956-57 petroleum crisis, the European na-
tions were forced into increased industrial activity.  Refinery capacity in the OEEC area increased from 127 million
tons in 1955 to some 180 million tons in 1959.  At the same
time, 16 of the 18 member countries were expected to possess their own refinery capacity within a very short time.
Not only had the number of the refineries in the industrial
countries grown, but their size was increasing as well.
Similarly, as a result of the number of transportation bottlenecks that had existed because of the closure of the
Suez Canal, a great number of orders for new tankers were
placed in the world's shipyards.  Between the end of 1955
and the end of 1959, the world tanker fleet rose by nearly
50 percent in terms of tonnage, thus far exceeding the volume of sea-borne oil, which rose only from 350 million tons
in 1955 to 470 million tons in 1959, the equivalent of a 34
percent increase.  The disparity between oil volume growth
and tanker tonnage increase was actually larger than represented by these numbers, as along with the new tanker orders

202

there came also a large number of technical improvements, including an increase in the speeds of the modern tankers, which meant that their actual carrying capacity was even greater than the tonnage figures indicated. Consequently, the overbuilding of new tankers caused a depression of tanker charter prices.[44]

The abundance of petroleum supplies, the expansion of refinery facilities, and the even larger increase of the tanker fleets made the years 1957-59 a buyer's, rather than a seller's, market.[45] The year 1960 was seen as a time when a new balance was achieved.

Out of Europe's total requirements, only a small portion of oil supply was contributed from indigenous production. But even here, Europe averaged a growth of about 9 percent a year, which, incidentally, coincided with the rate of growth in demand in the OEEC area. Thus, while European oil production (including shale oil) in 1955 was only 9,356,000 tons, it totaled, in 1959, over 14,811,000 tons (including 1,478,000 tons of Saharan production).[46] (See Table 19.) Evidently, these quantities could meet only a small fraction of European demand, most of which was being supplied by foreign sources. Due to Europe's rapid conversion, in many areas, from coal to oil, its economic dependence on those foreign sources increased, too. Arab sources--Algeria, Libya, and the Middle East oil-producing nations--thus held in their hands the key to Europe's prosperity. The balance of the rapid growth in demand was supplied, in steadily increasing quantities, by the Soviet Union.

In 1958, the common market for coal came into operation, and when, in 1960, a new balance had been achieved in the petroleum market, Europeans began to consider the various existing plans, and some new ones, for a unified energy policy for Western Europe. The main champions of such a policy were the original members of the European Coal and Steel Community, namely, France and Germany, which, together, represented "the third largest concentrated consuming area for hydrocarbons in the world, after the U.S.A. and U.S.S.R."[47] The French Government broached an all-European energy policy plan, and on April 21, 1964, the OEEC's Council of Ministers at Luxemburg accepted two of its recommendations:

1. Sources of supplies for oil should be diversified and should not depend upon one supplier or one single geographical area of supplies;

2. Importation of oil should be subject to certain controls.[48]

TABLE 19

Gross Imports of Crude and Products in OEEC Area

| | 1955 | | 1959 | |
|---|---|---|---|---|
| | Million Tons | Percent Share | Million Tons | Percent Share |
| Crude | | | | |
| Western Hemisphere | 10.9 | 11.3 | 18.5 | 13.3 |
| Middle East | 83.3 | 86.1 | 113.0 | 81.4 |
| Africa | 0.1 | 0.1 | 3.0 | 2.1 |
| USSR and Eastern Europe | 0.3 | 0.3 | 3.3 | 2.4 |
| Other | 2.1 | 2.2 | 1.1 | 0.8 |
| Total | 96.7 | 100 | 138.9 | 100 |
| Products | | | | |
| Western Hemisphere | 13.2 | 71.0 | 15.3 | 51.3 |
| Middle East | 2.5 | 13.4 | 5.8 | 19.5 |
| Africa | -- | -- | -- | -- |
| USSR and Eastern Europe | 2.4 | 12.9 | 6.9 | 23.1 |
| Other | 0.5 | 2.7 | 1.8 | 6.1 |
| Total | 18.6 | 100 | 29.8 | 100 |
| Crude and Products | | | | |
| Western Hemisphere | 24.1 | 20.9 | 33.8 | 20.0 |
| Middle East | 85.8 | 74.4 | 118.8 | 70.4 |
| Africa | 0.1 | 0.1 | 3.0 | 1.8 |
| USSR and Eastern Europe | 2.7 | 2.3 | 10.2 | 6.0 |
| Other | 2.6 | 2.3 | 2.9 | 1.8 |
| Total | 115.3 | 100 | 168.7 | 100 |

Source: OEEC, Oil: The Outlook for Europe (Paris, January 1961), p. 50.

Further, the Council of Ministers "agreed to agree" on a common energy policy for the OEEC nations within the boundaries set by the Treaty of Rome and the Treaty of Paris.[49]
   In June 1962, representatives of the High Authority Commission, the Common Market Commission, and the Euratom Commission formed a committee whose task was to draft proposals for a common market energy policy. This "interexecutive" committee turned in a set of proposals that covered nearly every aspect of oil purchasing, production, marketing, and trade, including such items as prices, tariffs,

and stockpiling.  The gist of the committee's plan was a
basic recommendation for a free and open petroleum common
market.  This was accepted, with some changes, by the Council of Ministers, and its first moderate application by the
European nations took place in 1965.[50]

But inasmuch as it was generally desired to arrive at
a mutually beneficial energy policy, it appeared that the
common good and the national interest did not, in all cases,
go hand in hand.  The NATO command was unable to form a unified policy to meet the external threat posed by the Soviet
oil trade challenge, and the OEEC Council of Ministers
seemed unable to agree on a common energy plan on the domestic scene.  Each nation seemed to pursue its national interest according to the availability of a diverse set of energy
resources in relation to its own particular requirements.
All these nations had constructed refineries to satisfy both
domestic demand and foreign trade needs.  Some European nations ventured into extensive offshore explorations, while
others began to invest heavily in regions that formerly
seemed to be reserved exclusively to the seven major international oil companies.  As cases in point where no common
grounds could be set up between the competing national policies were the question of the security of supply and the
problem of competition between energies.  According to Paul
Hatry,

> It is a common interest of enterprises and
> governments to seek short and long-run
> security of supply.  In the short-run, a
> common stockpiling scheme to meet about 65
> days of consumption was proposed by the
> Commission of the E.E.C. in 1964.  The
> fate of this proposition is most disappoint-
> ing as an example of the slowness of agree-
> ment in the Community on a purely technical
> proposal. . . .
> Finally, there seems to be a new bias
> in the European Commission towards favor-
> ing "European" oil companies as a factor
> to help security of supply and competition
> in the industry. . . .
> Competition between energies:  This
> seems to be the main problem in some E.E.C.
> countries.  Indeed, hostility not avowed,
> but hidden, towards oil, seems to be one of
> the main characteristics of some political-
> ly influenced men in European governments.[51]

How, in sum, then, did Western Europe meet the challenge of the Soviet petroleum offensive and the squeeze put on the Western European nations by the U.S. government to refrain from availing themselves of the low-cost oil that the Russians were offering them? What seemed to many Americans to be a major dilemma between national self-interest and the common good of the entire Atlantic community was quickly resolved by the Soviet Union itself, which in the mid-1960s began to limit its oil exports. Western Europe continued to buy oil and gas from the USSR, but in reduced quantities that did not seem to threaten its security.

At the same time, European governments were investing heavily in the expansion of overseas operations by their own national oil companies. French and Italian corporations were exploring for and producing oil in the Middle East and Africa alongside and in competition with the older U.S. and British companies. Their notable success encouraged their governments to develop national oil policies in support of these public enterprises, according to which an increasing ratio of the national economy depended on these new sources. This development, although motivated by the need to rid themselves of the utter dependence on Anglo-American oil, resulted also in the imposition of additional curbs on Soviet oil imports into Western Europe. Moreover, it increasingly tied the hands of those governments whose oil policies were nationalized to circumstances prevailing in the producing countries, over which they had no control. The narrow limits of available alternative resources turned these governments into political captives of their own economic policies.

## TANKERS ARE LARGER THAN EVER

The 1956-57 Suez Canal petroleum crisis drew the European nations' attention not only to new goals of oil production, refining, and marketing but also focused it on a rapid expansion of sea-borne transportation. Their achievements in this field stood in marked contrast to the sad state of the shipping industry in the United States. The success of foreign shipyards in constructing a large number of modern and improved tankers emphasized the inept handling of merchant shipping problems by Congress. While Congress was persuaded to allow the transfer of tankers to foreign flags on a two-to-one or a three-to-one basis as against new ones being built by the owners in U.S. shipyards,[52] the difficulties "in maintaining an adequate

American Merchant Marine and doing so within the bounds of
a free economy" were augmented by the obvious, attractive
economic advantages offered by the flags of convenience as
well as the serious resultant loss of jobs for U.S. craftsmen and seamen through such transfers. "The consequent
diminution in reservoir of experienced, qualified" manpower
could be "serious particularly from the standpoint of national defense."[53]  In the absence of a sufficiently reliable merchant marine, it was often suggested that in time
of emergency the United States could rely on its allies
among the European maritime nations.  But the navy came
out strongly against this notion.

> We should keep in mind that the interests
> of the United States are global, and emer-
> gencies may well arise wherein our inter-
> ests would not be identical with those of
> our European Allies.  Even within NATO
> framework it appears probable that the com-
> bined requirements of the NATO nations will
> exceed the total capabilities of the NATO
> shipping pool.[54]

In Europe, on the other hand, the 1956-57 crisis gave
great impetus to renewed shipbuilding activity.  New orders
were being made for larger and faster tankers.  Experiments
were being made with nuclear power for propulsion of these
merchant vessels.  The voyage of the nuclear-powered submarine _Nautilus_ under the North Pole ice cap in August 1958
demonstrated the possibility of a similar commercially
oriented nuclear-powered submarine tanker.
In 1959 the OEEC nations owned 55 percent of the total
world tanker fleet, which represented an increase of 11
percent for the area.  Greece alone increased its fleet by
71 percent, Liberia by 26 percent, Japan by 25 percent,
Sweden by 21 percent, and Panama by 10 percent.  At the
same time, the U.S. tanker fleet increased by 6 percent.[55]
Considering that the standard T-2 tanker was of
16,500 tons deadweight (dwt.), the construction of new
supertankers, "mammoth" tankers, "colossus" tankers, and
"gargantuan" tankers built in rapid succession during the
1960s to meet increasing petroleum requirements throughout
the world was a phenomenon of staggering proportions.
Larger and faster tankers kept rolling off the drydocks
into the seven oceans.  In December 1962, the British company Vickers-Armstrong (Shipbuilders) Ltd. received its
first order for a 100,000 ton dwt. tanker to be built for

the British Petroleum Tanker Co. Ltd., and designed to be
completed in 1964.  In that same year, three "mammoth" tankers were launched--Universe Apollo, Universe Daphne, and
Manhattan--each one of them of 130,050 tons dwt.  In January 1963, Japan launched the Missho Maru, which, at 132,334
tons dwt., was the largest cargo vessel ever constructed.[56]
     This rapid increase in tanker building continued to
grow from the 104,000 ton dwt. British Admiral, completed
in 1965, the largest British tanker so far, to French
Shell's three giant tankers--ordered in 1965--Magdala,
Miralda, and Myrtea, each one of 210,000 ton dwt.  The
same year Japan started the construction of the 205,000
ton dwt. tanker Tokyo Maru.[57]
     The six largest vessels on order in 1966 were ordered
in Japan by National Bulk Carriers for charter to Gulf Oil.
Each one of these tankers was 276,000 ton dwt. and 1,135
feet by 176 feet and had a depth of 105 feet and a draught
of 72 feet.[58]
     Capable of building increasingly larger tankers, ship-
yards moved from the 200,000 ton dwt. class of "mammoth"
tankers to the 300,000 ton dwt. "colossus" class and above.
In 1967 the Greek Goulandris group requested an estimate
for a 400,000 ton dwt. "gargantuan" tanker, while the British Lithgrows of Port Glasgow was tentatively exploring the
concept of an oil tanker of 500,000 tons dwt., with a length
of 1,370 feet and a draught of 85 feet.[59]
     As the size of tankers increased, the size of ship-
building berths in drydocks was also increasing to keep
pace with construction requirements.  There was no reason
to doubt the practicality of such ships.  Every move made
since the 1950s was in the direction of reducing the number
of smaller tankers.  The only limiting factor was the
draught, which, in the "gargantuan" tankers, posed a serious technical problem.  However, the growing use of offshore terminals soon alleviated the problems of loading and
discharge, while increased-size graving docks were being
built in the major ship construction centers of the world.
     This represented a most striking development in the
tanker industry over a relatively very short period of
time.  To appreciate the spectacular growth in tanker size
and speed, one only has to compare the 1967 "gargantuan"
class of tanker with the 1946 average tanker of 12,500 tons
dwt. and a speed of 12.85 knots.  According to Preston
Nibley and Alan Nelson,

          The reason for this trend is that larger
          tankers can move oil at lower cost than
          smaller tankers.  Larger tankers can be

operated by about the same number of men as
the smaller ships, reducing the cost of
labor per barrel transported. . . .  Fuel
consumption per barrel moved decreases as
the size of the tanker increases, even
though the larger and newer tankers generally are faster than the smaller and older
types.  Also, the investment per deadweight
ton for a larger tanker is much less than
that for a smaller ship, since the shipbuilding costs and the steel requirements
are relatively lower.  The reduced investment per unit of carrying capacity results
in lower depreciation, interest and insurance charges per barrel of oil transported.[60]

In light of this development in ocean transportation,
it is clear that even before the closure of the Suez Canal
in 1967, the canal's value as a waterway passage for the
new tanker fleets of the world had already become obsolete.
That branch of the general oil industry system that pro-
vides the sea-borne transportation had already determined
this fact.

## SUMMARY AND CONCLUSIONS

As a result of the Soviet oil offensive, concerned
oil industrialists attempted to enlist public opinion and
generate pressures on the U.S. government to use its political and military leverage with the consuming nations
of Western Europe in order to place curbs upon this trade.
The producing nations, upset because the Western oil companies found themselves impelled to meet the Soviet competition by lowering posted prices of Middle East oil, created an organization whose task was to limit the power of
these companies to make any more unilateral decisions in
the future in regard to the price of international oil.

While commercial considerations impelled oil industrialists and producing nations to oppose the incursions
of Soviet oil into noncommunist markets, the U.S. government was far more concerned with the political and military
consequences of this trade on the noncommunist nations.  In
the underdeveloped part of the world, dependence on Soviet
oil in exchange for indigenous staple goods seemed to create a long-range economic dependence of these nations upon
the eastern bloc.  This economic dependence, it was feared

in Washington, could easily be manipulated into a political alignment. At the same time, Washington officials were also viewing with growing misgivings the increasing East-West trade between Western Europe and the Soviet Union. They seemed to fear that these ties, if tightened, might place political as well as economic leverage in the hands of Moscow, by which the Soviet Union could attempt to weaken, if not entirely divide, the NATO alliance.

U.S. apprehensions did not seem to be equally shared by the Western European nations, which continued to trade with the Soviet Union. That Soviet oil did not become the major import into these countries that it could have was due, primarily, to Soviet shift of its foreign oil policy after Khrushchev's fall from power. Trade in other commodities, especially Western European heavy machinery and industrial equipment, continued to grow under the new leadership at the Kremlin. However, in matters of oil, Europe, since the crisis of 1956-57, strove to become as independent of external influences as possible, thus nullifying both Soviet and Anglo-American pressures. Western European achievements in the field of foreign explorations and deep-sea drillings for oil and gas are as remarkable as the rapid growth of its domestic petrochemical industry. Europe had made giant strides since World War II, and in oil and chemicals, as in many other industrial fields, it has become a rival of U.S. leadership in world markets. The swift ascendance of Europe has caused many a U.S. decision-maker to pause and rethink the past course of the Western alliance and the inherent direction of the Atlantic community. The lack of economic imagination and the absence of U.S. confidence in a quick European recovery were, in 1948, at the basis of the present shifting disparity in the Atlantic partnership.

The concept of the oil industry as it developed in the 1960s would not be complete if we did not include, side by side with the story of production and manufacturing, the trend of the tanker industry, which is an equal arm of the whole integrated oil system. Thus, the rapid increase in the size of tankers during those years brought up the pertinent question of the relevance of the Suez Canal to present-day oil trade. This question will be discussed in greater detail in the following pages.

# 10

In 1967 the Suez Canal was closed to all sea-borne transportation.  As far as the Persian Gulf and Saudi Arabian oil industries were concerned, the Red Sea was now a dead-end waterway and could serve their purposes no longer. At the closure of the canal, memories of the 1956-57 crisis immediately surfaced to put into motion a similar set of precautionary measures as were employed in the previous decade.  On June 13, 1967, one week after the beginning of hostilities in the Sinai Desert, the Middle East Emergency Committee (MEEC) convened to draft a plan of action, similar to the one that had been put into operation 11 years earlier.  Most significantly for the development of the oil industry in the past 20 years, in respect to the diversification of production, the enlargement in size of the crude oil tankers, and the construction of a self-sufficient refinery industry in the consuming countries, it was soon realized by everyone concerned that no real emergency existed.  Therefore, on August 15, 1967, the Foreign Petroleum Supply Committee (FPSC) declared that no emergency schedules were needed.[1]  Thus, the closure of the Suez Canal in 1967 made only a slight ripple in comparison to the 1956 affair. This extraordinary shift was due to the major changes that not only had altered the entire structure of the world petroleum industry but also, in the process, had affected the direction and purposes of U.S. foreign policy in that region.  Weighed by the Department of State on one side of the scales were Western petroleum requirements, produced by Western companies for Western energy consumption.  On the other side were the rapidly sliding balance of power and a new set of equilibrium probabilities.

During the 1950s, a new political complex had emerged in the Middle East.  The great added single factor introduced

into the area, from the point of view of U.S. foreign policy, was the rapid increase of Soviet influence. The high
priority this region enjoyed in U.S. strategic policy, and
the threat to the area by the presence of a powerful and
hostile rival, dictated to Washington officials that a set
of drastic changes be made on several levels. In relation
to oil, these revisions of long-standing policies were
largely evident on two plateaus: (1) U.S. relations with
the international petroleum companies, and (2) U.S. relations to Western Europe's petroleum requirements. The
following two sections of this chapter deal with these two
distinct aspects of U.S. foreign oil policy as they were
viewed by many U.S. government officials in 1967.

## U.S. RELATIONS WITH THE INTERNATIONAL PETROLEUM COMPANIES

As long as Great Britain maintained the security of
Western interests in the Middle East, Washington officials
were satisfied to stay out of the political and military
picture of that region. Both the U.S. government and the
American oil companies, however, believed that their respective interests in these countries would be better
served if U.S. money and know-how would be invested in and
directed toward the general welfare of the host nations.
This intention was to be chiefly realized through material
contributions aimed at raising these nations' standard of
living. U.S. companies invested large sums of money out
of their profits in order to train indigenous skilled labor and to build schools and roads and modern housing for
the personnel. Modeled after the middle-class suburb in
the United States, small towns sprouted in the deserts near
oil wells and refineries. Progressive classes for the
training of skilled labor were held continuously by teams
of experts, and many promising young men received substantial grants to pursue their oil studies at some of the best
technological institutes in the United States. In the eyes
of the companies as well as many Washington officials,
these activities constituted a private foreign aid program
that ought, according to contemporary reasoning, to contribute heavily toward cementing lasting ties of friendship
with the recipient nations.[2]
This self-congratulatory self-image of the companies
clashed sharply with the prevailing mainstreams of Arab
public opinion. Nothing could possibly demonstrate better
to Arab masses, who had never seen the Western world, the

immense gap between their own standard of living and all
the material wealth demonstrated in the companies' local
aid programs.  The modern-looking housing developments,
equipped with electric appliances and air conditioning,
and surrounded by flower beds, green lawns, and clean
paved streets, provided a sharp and unpleasant contrast to
the general backdrop of poverty, shack dwellings, and dung-
strewn earthen roads.  No amount of companies' generosity
could divert the tide of mounting resentment against these
foreign enterprises, which seemed to prosper so well on
the exploitation of the Arabs' own national resources.
The close cooperation that had existed between the U.S.
government and foreign oil interests since the 1920s made
U.S. foreign policy itself suspect of conspiracy.  There
was a vociferous segment of Arab public opinion that viewed
U.S. Middle East policy as a pawn in the hands of the international oil cartel by means of which the latter fur-
thered its own interests in obtaining large financial gains
through the unchecked exploitation of indigenous resources.
This view was in accord with Marxist-Leninist tenets, which
attributed to corporate monopolies the ruling-class status
in the overall U.S. power structure.  Another stream of
Arab public opinion, however, more committed to Western
frame of references, in reverse process of political analysis, assigned the private oil companies the role of tools
in the hands of the U.S. government by means of which that
government was able to gain a foothold in the Middle East.
This view enlisted the political theory of power politics
on its side.  Both views, the socialist and the nationalist, created massive antagonism to the prevailing concessionary policies.[3]  The words "concession" and "concessionary rights" became anathema to both socialists and nationalists in these countries.

As long as only Western powers were present in the
Middle East, U.S. foreign policy could continue to back
the economic interests of the foreign oil companies by in-
tervening on their behalf wherever the need arose.  But as
soon as Soviet policy became active in the region, this
line of action became outdated.  Soviet presence affected
the political status of these companies in the producing
countries as it shrewdly played upon the widely spread
sentiments of nationalism and anti-Westernism.  In response
to this surging alignment between Soviet power and local
political movements, it became necessary for official U.S.
policy to moderate its links with the interests of the private oil companies in the Middle East.  Good official policy, then, required the U.S. government to disassociate

itself from taking sides in any future disputes that might
arise between governments of the producing nations and
their contractual oil concessionaries.

With the passage of time the Department of State
seemed to keep more and more aloof from foreign oil policies.  Official communication was narrowed to a rigorously
codified form of exchange.  The actual point of contact between government and companies occurred in the jurisdictional scope of the Department of the Interior.  Assigned
the task of looking after the domestic requirements of the
nation and motivated also by antitrust considerations, the
Department of the Interior became the initiator and coordinator of government-industry relations in respect to U.S.
foreign petroleum policy.  Pressed even further by a series
of oil crises in the Middle East, the Department of the Interior created the FPSC to cooperate with the international
corporations, principally in time of emergency, in the redistribution and direction of movements of crude oil and
oil products in the free world.  This organizational structure within a government department that had no jurisdictional powers over foreign affairs decision-making was in
complete accord with the prevailing notion that the less
visible the ties with foreign oil concessionaries, the better the U.S. image in those countries.

While the U.S. government was thus shifting its official position in relation to foreign oil concerns, there
was also a marked reversal being made in the companies'
policy with respect to their relations with the U.S. government.  The insecurity of their investments in the Arab
countries--demonstrated by several cases of nationalization
of property, confiscation of assets, and the general trend
toward modification and even cancellation of contracts--
prompted the companies to alter their stance in respect to
government authority.  During the 1960s, it frequently became necessary for oil companies to turn to the U.S. government for assistance in their disputes with foreign governments.  Such assistance as was requested was not forthcoming.  In an increasing number of cases, the Department
of State subordinated its traditional role of protecting
the interests of U.S. concerns abroad to the overwhelming
policy imperative, created by existing conditions in a
polarized world, of making friends with foreign governments
and influencing other peoples.

While U.S. foreign policy was moving away from foreign
oil considerations in an attempt to disassociate itself
from any concessionary vestiges of a bygone era, the oil
companies were in fact becoming increasingly dependent on

the good will of State Department policy-makers.  The need
for security of investment, which was not being satisfied
by government policies, prompted the international oil companies to look for safer and friendlier lands.  New oil exploration ventures soon began to take place in areas that
up until that time had been passed over because these explorations were much more costly than those done in Middle
East countries, where every fifth drilling reportedly
turned into a producing well and where oil was found very
close to the ground surface.  Nevertheless, the oil companies seemed to prefer the safety of their future investments and the guarantee of their continuance as against the
cheaper costs but unrealiable contracts of Arab nations.
Thus, new operations spread throughout Canada and Alaska,
Nigeria and Southeast Asia, in offshore waters, and in the
deep ocean beds.

      These new directions of the international oil corpora-
tions were in full accord with U.S. foreign oil policy as
enunciated by President Eisenhower after the Suez crisis
and adhered to by three successive presidents.  The heart
of this policy in regard to petroleum was the diversification of the sources of energy and of their geographical
locations as a precaution against an exclusive dependence
on one type of energy located in a single part of the world.
In order to promote this policy, many companies were encouraged materially by the government to go into new fields of
exploration.  These new vistas of opportunity and the immense prospective gains promised attracted the major companies as well as a swelling number of independents, which,
in the 1960s, competed with the majors successfully for
production and markets.  Especially successful among the
new independent companies expanding into foreign territories of exploration, transportation, shipbuilding, and refining were the Occidental Oil Company, the Continental
Oil Company, and the Marathon Oil Company.


                U.S. RELATIONSHIP TO WESTERN EUROPE'S
                        ENERGY REQUIREMENTS

      Much of U.S. foreign policy concerns the continuing
survival of a Western European community of free nations
able to contribute materially to the NATO alliance and international prosperity.  In order to maintain this alliance, U.S. administrations will subordinate nearly all alternative policies.  It is a _sine qua non_ of this policy
that the flow of energy into the Western European community

must not be interrupted. To accomplish that aim, under the most adverse conditions, the United States enlisted an enormous machinery of government employees, cooperated with private oil corporations under penalty of violating antitrust regulations against monopolistic and price-fixing practices, and directed a worldwide reorganization of production, refining, and transportation never before attempted on such a grand scale except, perhaps, during World War II.

At the same time, the rapid growth of the Western European economy has steadily increased its demands for greater sources of energy and thereby spurred private companies as well as governments to make substantial investments in the development of the foreign oil industry and all its ancillary branches. As a result of these efforts, refineries and petrochemical plants throughout Western Europe were gradually fulfilling that continent's requirements for petroleum products and exporting large quantities to other nations.

Crises made this remarkable expansion possible. An acute crisis in dollar reserves after World War II provided the initial drive toward a self-sufficient Western European refining and petrochemical industry, and continuing political and military crises in the Middle East during the following two decades spurred the effort toward higher goals.

Having a self-sufficient refining and petrochemical industry, Western Europe's fuel shipments from the Middle East have increasingly consisted of crude oil. As crude requires larger bulk containers than its products and as Western Europe's energy demands have been steadily augmented from year to year, tankers had to be built that became progressively bigger and capable of holding larger quantities of oil. As soon as the average tanker tonnage had surpassed the 60,000 deadweight ton mark, it became obvious that the new tankers already on order in the leading shipyards of the world would be unable to use the Suez Canal. Therefore, efforts were successfully expended on enhancing the speed rate of all tankers above the 100,000 deadweight ton mark (which would be compelled, because of their size, to use the round-the-Cape route) in order to reduce the time difference usually added to this voyage as compared with the shorter route through the Suez Canal.*

---

*"Vessels of more than 65,000 tons cannot use the Suez Canal, but even taking the longer route, a 150,000 to 300,000-ton tanker can compete successfully with

Supertankers, giant tankers, mammoth tankers, and gargantuan tankers have built up speeds that have decreased this
time difference, in some cases, to less than six hours.
Having thus neutralized the time element between the two
routes, and at the same time being able to accommodate much
more freight, these tankers' trip around the Cape of Good
Hope became more lucrative to the oil companies than the
shorter canal route.  The profit margin increased even more
as soon as West Africa became an important importing and
exporting market.  Special docking and pumping facilities
were constructed in South Africa and along the Atlantic
coast of West Africa in order to accommodate outsized tankers.  A repair dock in South Africa rapidly replaced Malta
as a major stopover for all ships and tankers traveling between South Asia or Arabia and Western Europe.  The discovery of oil in Nigeria added to the versatility of tanker
loads.  Because of their gigantic size and because of many
recent discoveries of petroleum resources in West and North
Africa as well as the construction of refineries in many
new producing and consuming locations, tankers now became
compartmentalized.  This revolutionary innovation in tanker
construction contributed to yet another increase in profit
margins.  While the small tankers, which could carry only
one type of bulk, usually delivered to a single destination, had to return to their home port in the Middle East
on ballast, the compartmentalization of the larger tankers
made it possible for shipowners to diversify their carrying loads as well as their destinations, thus avoiding the
waste of empty container trips, which doubled the time element spent on each shipment as well as the expense involved.
Thus, the emergence of the oversized tanker contributed to
greater flexibility and profits in the petroleum industry.
The expenses of maintaining a larger tanker do not far exceed those spent on a smaller one.  The construction cost
as well as the insurance premium are reduced as per ton;
the crew of either type of tanker consists of 41 men and
the amount of fuel required to operate a larger tanker is,
proportionately, far less than that consumed by a smaller
tanker.
    By the time of the final closure of the Suez Canal in
1967, the petroleum industry had already lost all practical
interest in this waterway.  Since the Mediterranean nations

---

smaller, short-run tankers for cargoes destined for European ports along the Atlantic" (Ralph E. Williams, "Strategy and Oil," Naval Review, 1967, p. 157).

derived their Middle East oil supplies by pipelines that
ran from wells to pumping facilities along the eastern
shore of that sea and since, moreover, new oil and gas resources were being continually discovered in the Mediterranean regions* as well as in Nigeria and the Northern Sea,
the Suez Canal rapidly became a highly dispensable asset
as far as oil transportation was concerned. The European
nations no longer needed to rely on its continuous operation, as they had in 1956-57. No armed conflict need occur
to reopen the canal for Western economic interests. The
Suez Canal has become a noncommercial asset and should be
evaluated only in terms of its political and strategic
properties.

Not only commercially but also strategically, the
round-the-Cape voyage appealed to policy-makers. The Atlantic Ocean is a "Western" waterway. It is the indispensable bridge for the partnership that constitutes the Atlantic alliance. It is a protected ocean, controlled by
NATO's naval and air forces and uncrowded by hostile fleet
movements. It is a safe route for a highly strategic commodity such as petroleum. To be able to transport such
strategic oil to all parts of the world, the United States
must be able to rely on a worldwide network of waterways
that are militarily and strategically secure. No matter
what military policies may evolve over the years in response to a wide range of political and technological vari-
ables, at the heart of all these policies there will always be the concept of an oceanic doctrine. If we judge
future policies by this one constant factor--the wide open
oceans--the route that petroleum transportation is now
making around the Cape of Good Hope seems to be preferable
to the Mediterranean trap.


SUMMARY AND CONCLUSIONS

In the 10 years between the time when the Eisenhower
Doctrine was first enunciated and the closure of the Suez
Canal in 1967, U.S. foreign policy in the Middle East had
achieved a remarkable freedom from commercial oil considerations. The expansion of Western European government-
owned oil concerns into areas formerly controlled by a
cartel of seven major international corporations and the

---

*These Mediterranean regions include, primarily,
Libya, the Algerian Sahara, Egypt, and the Po Valley in
Italy.

ownership by these governments or their citizens of the
world's largest tanker fleets as well as a viable and prosperous refining and petrochemical industry relieved the
United States of a great measure of responsibility for the
continuous petroleum flow into Europe. But the United
States still carried the full burden of responsibility for
political and military policies in the Middle East. As a
result, having isolated oil economics, U.S. foreign policy
is now able, to a degree, to deal with problems arising
from the political and strategic properties of oil, unen-
cumbered by "residual" commercial interests.

This division of labor, however, is not without its
limitations, stemming, mainly, from two major dilemmas of
interdependence. In the first place, U.S. demand for foreign sources of oil has been increasing steadily since the
1960s. Although most of these demands are being supplied
from Western Hemisphere sources, Middle East oil still
plays a role in present and future calculations. In 1967
about 3 percent of total U.S. petroleum consumption was
imported from the Arab nations. However, current quantities as well as future prospects fluctuated from nation to
nation. At the top of this list, calculated on the rate
of current production, stood Saudi Arabia, with proven reserves for the next 75 years; at the bottom stood Kuwait,
with proven reserves sufficing only for the next 12 years;
Libya followed with an estimated period of 17 years, while
Iran and Iraq occupied an intermediate position of 55 and
35 years respectively.* While this list reflects the
scale of U.S. economic interest in the natural resources
of these nations, it does not necessarily agree with the
scale of U.S. strategic concerns in the Middle East. While
Saudi Arabia is of secondary strategic importance at present, Iran and Iraq, no doubt, top the list of preferred
military territories. At the same time, Kuwait, with its
present large supply of oil, has barely any strategic significance in the current scheme and, with the exhaustion
of its oil supplies, will most likely be reduced to the
role played nowadays by Jordan. On the other hand, a non-
oil-producing country like Lebanon can be an enormous geopolitical asset to the Western alliance.

The second dilemma of overlapping but unparallel advantages concerns the commercial and economic interests of
Western European nations in preserving the continuous flow

---

*This view of Iraq's future prospects is about to
change soon due to reportedly recent new discoveries of
oil resources, the extent of which is still kept secret.

219

of Middle East oil to their ports.  While these nations
seem to prefer to relegate the political and military responsibility for countering increasing Soviet influence in
this region to the United States, they insist that such
policies as are necessary to meet that threat must not conflict with their own commercial interests.  In this conflict of interests between U.S. government policies and
Western Europe's requirements, the international petroleum
industry strongly identifies itself with European consuming interests.  At the same time, it also keeps insisting
on increasing U.S. government political protection from
antagonistic indigenous measures often attributed to hostile Soviet sources.

However, in 1967 the general problem of this precarious balance of interests seemed to be somewhat alleviated
by the discovery (but insufficient development) of complementary sources of energy.  Petroleum explorations in such
far-flung areas as Alaska, Canada, Indonesia, and Nigeria
were supplemented by new resources of natural gas from the
Sahara and the Northern Sea.  In the United States, the Department of the Interior, through its Bureau of Mines, is
promoting the development of shale oil deposits and is per-
fecting new technological methods to extract a larger ratio
of oil from each producing well.  It is also engaged in
liquefying and gasifying processes of indigenous coal resources.  Oil explorations now expand widely beyond the
territorial boundaries of the United States deep into the
continental shelf, an expansion that in itself has been
directly responsible for much of the recent modifications
and amendments of legislation concerning territorial waters
as well as fisheries.  In the 1960s, interest was steadily
growing in undersea resources of the ocean bed as was expressed during the geophysical year initiated by the U.S.
government and in a series of related conferences undertaken by United Nations committees.

In 1967, then, a general confidence existed in many
circles in the abundance of world energy resources.  When
the Suez Canal was closed to all sea-borne transportation,
when pipelines were being sabotaged and international companies were being nationalized, the industrialized world
did not panic and no military expeditions were started.
The existence not only of more profitable routes of transportation but also alternative sources of energy seemed to
be confirmed by almost daily discoveries of new producing
wells in all parts of the world.

The periodical scares about the imminent scarcity of
petroleum resources as well as the dismal record of the

petroleum industry regarding the general environment prompted many government officials and legislators to look into alternative modes of energy. The development of industrial atomic energy, which had been gravely handicapped by problems of residue storage and radiation, seemed to be nearing a solution to the safety problem. Attempts were also being made to develop the electrically energized car, which some experts believed could be perfected by 1975. Were that done, the petroleum industry would lose its largest customer--the automotive vehicle. At the same time, this shift in the motor mechanism of the car could be a significant contribution to efforts to clean up the polluted environment. Another source of energy already vying for recognition in its own right and as an ecological improvement was the solar energy system periodically discussed by Congress.

Commercially, then, the changes that were wrought upon the petroleum industry by the closure of the Suez Canal in 1967 and the increasing influence of the Soviet Union in the Middle East seemed at that time to be met successfully by an apparent worldwide diversification of explorations and energizing resources. This complacency, however, in regard to the security of supply of Western petroleum requirements, in general, and the near self-sufficiency of U.S. resources was shaken by the energy crisis of the early 1970s and the emergence of Middle East oil as a political weapon against the United States.

# 11

## THE PROHIBITIVE
## COST OF ARAB OIL

In 1972 and 1973 the United States was abruptly awak-
ened from its euphoric oil slumber.  The prevailing faith in
the great abundance of oil resources in general, and the
near self-sufficiency of the domestic industry in particular,
to supply all national energy requirements, seemed to be
shattered by a series of nationwide shortages.  The much-
publicized doctrine of the diversification of the geographical locations and of the forms of energy resources, which
had been accepted as the cornerstone of oil policy since the
Suez Canal petroleum crisis of 1956-57, had not been sufficiently applied in reality to be able to prevent emergency
situations of sudden shortages, actual or otherwise.  Both
government and industry bore the responsibility for this
gross mishandling of the public welfare.

Government agencies were prevented from actuating such
programs by judiciary and industry opposition to public intervention in a traditional area of private enterprise; such
opposition was based on the fear of government control over,
if not complete takeover of, the entire U.S. oil industry.
Thus, old fears of the concentration of great economic power
in the hands of a few federal agencies placed curbs upon
government oil planning.  Government negligence in its role
of encouraging the policy of energy exploration and diversification was further augmented by the general belief, held
during the 1960s, in the ready availability of large amounts
of such resources in the Western Hemisphere.

At the same time, the U.S. oil industry, operating both
at home and abroad, was reluctant to take such financial
risks and make investments that could not be counted on to
bring in assured returns.  Based on the commercial premise
of the economics of cost/profit calculations and being

responsible to its stockholders, the private sector was not
sufficiently motivated to redirect enough of its exploration
efforts into sometimes remote experimental fields.

This final chapter attempts to abstract some of the
major policy-making issues in the current oil controversy.
The first section presents the relevant facts in the controversy:  First, it explores seven causes generally given
for the present "oil crisis," and, next, it reviews the government's actions to alleviate this "crisis."  The second
section deals with the producing nations and the effect of
their politics upon international oil prices and policies.
The third section describes current developments of U.S.
tanker fleets and offshore terminal capacities in relation
to world availabilities in these areas and Arab oil trade
policies.  The last section analyzes the political and strategic uncertainties of the region in respect to Soviet presence and the wavering regimes and questions the wisdom of
entrusting the U.S. economy to the latter's hands.

THE 1972-73 OIL CONTROVERSY:
A GENERAL BACKGROUND

As the winter of 1972-73 approached, it became apparent
that the domestic oil industry and refineries were falling
behind in fulfilling standing orders for industrial and
heating requirements.  This progressively worsening situation culminated in particularly acute shortages in five midwestern states.  In the controversy that ensued, the private
sector laid the blame for the "crisis" upon the government,
while the government blamed the private sector for it.  In
the private sector, the majors and independents, the oil
producers and oil refiners, and the oil and gas industries,
exchanged mutual recriminations.  At the same time, the Office of Oil and Gas, the Office of Emergency Preparedness
(formerly the Office of Defense Mobilization), and the Congress seemed to be as eager to shift the balance of responsibility to each other.

On the surface, as soon as the shortages became apparent and before the barrage of accusations of malicious intent and gross negligence broke loose, there was a tendency
to explain the "oil crisis" on the grounds of seven major
causes.

In the first place, it was argued, the domestic refineries did not expand sufficiently to meet the rising national requirements.  Therefore, even if crude were available, it would have had nowhere to go because of the

inhibited refinery expansion. This argument was invalidated
as soon as it was discovered that many fuel oil refineries,
in spite of the evident increasing demand, were not working
at full capacity.

Next, the blame for the refiners' failure to fulfill
oil orders was placed at the doorstep of the gas industry.
The low ceiling enforced by law on the price of natural gas
for interstate trade made oil purchases seem like a losing
proposition in comparison. Thus, a large number of orders
were made for natural gas and not quite enough for oil.
This "repressive" price regulation of natural gas, and the
ensuing imbalanced market that resulted from it, caused some
striking anomalies in the distribution of oil. While Oklahoma, for example, enjoyed an overabundance of oil, its
neighbor Iowa was suffering from acute shortages.

Heating fuels were especially hit by inflation. Phase
Two of the Nixon Administration's economic planning placed a
freeze on heating fuels, but not on other types of oil products. Consequently, oil refineries stood to gain larger
profits by producing all but heating fuels. Thus, consuming states that depended exclusively on the domestic industry to supply their needs became the major victims of the
market imbalance.

However, the private sector was reluctant to shoulder
the blame for the "crisis." The oil industry, the gas industry, and the refiners were quick to point out that the
shortages were due indeed not to the marketplace but rather
to insufficient and short-sighted government policies that
curbed incentives for exploration of new oil resources.

Not only the government but nature too received its
share of the blame. The previous year (1971-72), the private industry argued, it had prepared large stocks of heating fuels, which it was unable to sell due to unusually warm
winter weather in the midwestern states, and thus suffered
considerable losses. This year (1972-73), the industry was
therefore, of necessity, more cautious and produced less
heating fuel oil. The unusually cold weather of this winter
found the industry, understandably under these circumstances,
quite unprepared to fill the great number of unforeseen orders that came in.

Another constant factor always taken into consideration
as a contributing cause to the "oil crisis" is an allegedly
excessive concern for the environment that gives little due
regard to industrial imperatives. Thus, new air pollution
laws have induced many communities to switch demand to fuels
of low sulphur content. At the same time, these new air
pollution laws have also caused a dislocation of industry

from its former centers of distribution to new sites and thus disrupted "traditional demand/supply trends."

Last but nevertheless not necessarily infrequently heard during the early days of the "oil crisis" was the argument about Pentagon spending. It was argued that as a result of the unforeseen resumption of large-scale bombing missions in Vietnam and Cambodia, the Defense Department had miscalculated its jet fuel requirements and was forced to place a new order for an additional 315 million gallons for the last six budget months. The Pentagon replied that such purchases as were made, first, did not contribute to heating fuel deficiencies and, second, were themselves badly hindered by the rise of jet fuel prices by 30 to 40 percent. As a result of this price hike in the marketplace, the Defense Fuel Supply Center was able to supply only 60 percent of its requirements, as many regular suppliers did not bid at all. Moreover, the Pentagon answered its critics, the real demand for all oil products for defense purposes was in fact the lowest in 13 years.

After one reviews the seven frequently given major causes for the oil shortages in 1972-73, two facts become glaringly evident. First, in no case was there a serious allegation made that the nation's oil resources were running out. As late as 1971-72, both the Office of Oil and Gas and the National Petroleum Council agreed with the long-established fact that there is "evidence of the high petroleum potential of the United States,"[1] that the supply, if fully exploited, would last well into the 21st century, and that other forms of energy could be perfected at commercial prices long before the petroleum supply ran out. There was no retraction of this statement at any time during the oil panic of the following winter. The NPC, however, stated that "the trend in the last decade of devoting a declining percentage of producing revenue to finding and developing production of crude oil and natural gas has resulted in a drastic decline in exploratory and development drilling which . . . is inimical to the development of the country's enormous petroleum resources."[2] The NPC further warned that "to the extent that policies of industry and government militate against accelerated exploration, particularly drilling, a high percentage of the petroleum resources of the United States is immobilized."[3] Insufficient exploration and drilling, and not the lack of availability of oil resources, was, then, at the bottom of the recent scarcity of crude.

The second fact that emerges from an examination of the seven major causes for the "oil crisis" is the gross negligence of U.S. conservation requirements. Since the 1930s, U.S. oil conservation policy strove to achieve underground reserve stocks sufficient to supply civilian and military demand in time of national emergency for a period of not less than 10 to 12 years. Such reserves were often drawn upon in the past in times of national emergency, such as World War II and the Korean War, and in times of particular hardship to U.S. allies in Western Europe due to oil shortages, such as the Iranian oil crisis of the early 1950s and the Suez Canal oil crisis of 1956-57. This last crisis drained U.S. oil reserves to the lowest level they had ever reached. These reserves were never refilled. Since that crisis, the domestic industry has been producing oil as close to the level of demand as the traffic would permit. Thus, the slightest unforeseen fluctuation in market demand could upset the whole balance of available supplies. In the absence of a ready stock of reserves shielded by national conservation laws, there was no available source to draw upon when shortages occurred in 1972-73 for the first time since the United States shipped its entire emergency reserve stocks to Europe in 1956-57.

The lack of sufficient exploration and the negligence of conservation, however, could be corrected only in the long range. The "crisis" of 1972-73 required immediate solution. Since no domestic resources were available to alleviate the shortages, a situation was quickly created where the wisdom of maintaining oil import controls was being challenged. The same taxes and quotas that had been initially placed upon foreign oil in order to protect the domestic industry from competition by "cheap" Arab oil as well as to protect the nation from overdependence on remote and uncertain sources of energy were now being repealed one by one by the Nixon Administration in order to meet a crisis situation with which it was ill-prepared to cope.

The restrictions imposed by President Eisenhower in his proclamations of 1958 and 1959 on the importation of foreign oil in general and Eastern Hemisphere oil in particular were gradually lifted by Nixon. As early as mid-May 1972, Nixon allowed an increase of crude oil imports to Districts I-IV* from 935,000 barrels per day to 1,165,000 b/d. These additional 230,000 b/d represented an immediate increase of over 14 percent. At the same time, Nixon approved the increase

---

*The United States is divided into five "oil districts": Districts I through IV cover the area east of the Rockies; District V covers the area west of the Rockies.

of Canadian oil imports from 540,000 b/d to 570,000 b/d,
representing an increase of only 5.5 percent.

These measures were hardly adequate to counter any unforeseen emergency situation.  The following winter (1972-73)
reported increase of overall demand rose by 6.5 percent,
while domestic new supplies increased only 5.3 percent.
Government import controls were further relaxed as a cold
winter was being forecast for the midwestern states.  Oil
imports into Districts I-IV were now raised a second time
from 1,165,000 in May to 1.8 million b/d in the fall.
Nevertheless, the National Oil Jobbers Council pressed for
suspension of all government controls over oil imports and
price ceilings on fuel oil.  In January 1973,[4] Nixon temporarily suspended controls on the importation of No. 2 fuels
(heating and diesel oils).  The suspension was to lapse on
April 30.  At the same time, the president raised imports
of crude oils for the third time in nine months from 1.8
million b/d to 2.7 million b/d in Districts I-IV and from
717,000 to 800,000 b/d in District V.  Imports of oil from
Canada were also somewhat raised.  However, it seemed that
in principle the president still adhered to the policy of
import controls as it had been enunciated by Eisenhower.
This policy was to stay in force in spite of the temporary
relaxation of such laws.  It was Nixon's intention to alleviate an immediate crisis situation by a short-term suspension of certain import controls while, at the same time,
working to fill the growing gap between supply and demand
through an adequate development of domestic resources.  Nevertheless, although increases in crude oil quotas and suspension of No. 2 oil controls had been purported to alleviate winter hardships and were intended to be discontinued
on April 30, further modifications of oil imports regulations were authorized by the president in April.  In a second proclamation within a period of three months, the president modified again the original Proclamation No. 3279,[5]
of the Eisenhower Administration and suspended all oil import controls.

The much-awaited oil policy statement announced by the
president in his State of the Union message to Congress on
February 2, 1973, was delivered on April 18.  It was followed immediately by a recorded statement for radio and tele
vision.  Nixon's second modification of Eisenhower's oil imports proclamation and his energy policy statement were made
simultaneously.  The public statement and the message to
Congress were used as means by which the administration
sought to justify the temporary removal of controls on oil
imports and, at the same time, to restate the administration

continued steadfast support of the national oil policy of
self-reliance on domestic supplies of energy.

> In order to avert a short-term fuel shortage
> and keep fuel costs as low as possible, it
> will be necessary to increase fuel imports.
> At the same time, in order to reduce our
> long-term reliance on imports, we must en-
> courage the exploration and development of
> our domestic oil and construction of refin-
> eries to process it.[6]

This new development of the domestic industry was to
be done in full accord with demands of national security and
the control of the environment. The president's message to
Congress thus contained six guideline principles for a long-
range oil policy for the development and expansion of all
domestic energy resources, both natural and derivative.

But, at the same time, in his message to Congress,
Nixon also proposed a program for the construction of deep-
water ports and offshore terminals to receive short-term oil
imports and the buildup of a huge fleet of supertankers in
which to carry these imports home. The enormity of this
last program and the skyrocketing expenditure that it would
entail present a strong argument to doubt the sincerity of
those of its sponsors who assert their intentions to maintain it as a stopgap measure by which to alleviate temporary
oil shortages. The magnitude of this program and the price
tag it carries make no sense at all in short run. It can
only be justified if it is intended to be used as a vehicle
of long-range U.S. dependence on Middle East oil. If this
is the case, and if this plan is truly intended to lead the
United States away and astray from announced public policies
of self-reliance in matters of energy, it is also bound to
lead the country on the downhill road toward a permanent
state of insolvency.

Moreover, the view that U.S. Middle East oil imports
should be regarded as a long-range program rather than a
makeshift expedient to solve temporary shortages finds an
added support in Nixon's reference in his message to Congress regarding an organization of oil-consuming nations.
Again, such a worldwide plan, requiring prolonged and extensive diplomatic efforts, can only be understood correctly
if posed against a permanent state of reliance on Middle
East oil imports.

However, in evaluating actual U.S. needs for Middle
East oil imports, we cannot close our eyes to extraneous

developments.  The most decisive of these independent factors
is that the Arab nations, through their separate governments
as well as through their Organization of Arab Petroleum Exporting Countries (OAPEC), are now in a very good position
to assert exclusive monopolistic controls over all crude oil
products originating in Arab wells, fix world market prices,
and even dictate flag preferences for the transportation of
their exports.

## OIL POLICIES OF THE ORGANIZATION OF
## ARAB PETROLEUM EXPORTING COUNTRIES

While U.S. dependence on Arab oil imports seems to have
increased in the short range, the Arab oil-producing states
have been promulgating restrictive oil policies that make
sense only if based on an indefinite U.S. dependence on Arab
resources as well as an infinite Arab capacity to supply
U.S. demand.  Since neither prospect is true, Arab pressures
to exploit a momentarily difficult situation in the United
States may conceivably boomerang at the oil-exporting nations
inasmuch as it will spur U.S. initiative to precipitate the
development of a domestic energy industry and thereby relieve
the economy, even sooner than earlier planned, of an unreliable business association.

The principal agents for Middle East oil resources are
the once-powerful international companies.  During the 1960s
these companies increased their oil production in the Eastern
Hemisphere from 5,690,000 b/d in 1961 to 18,020,000 b/d in
1971, as compared to an increase of the oil production in the
Western Hemisphere during the same period from 3,130,000 b/d
in 1961 to only 4,190,000 b/d in 1971.  (See Table 20.)

TABLE 20

World Oil Production by U.S. International Oil Companies
(barrels a day)

| Year | Western Hemisphere | Eastern Hemisphere |
|------|--------------------|--------------------|
| 1961 | 3,130,000 | 5,690,000 |
| 1964 | ↑ | 8,000,000 |
| 1968 |   | 12,000,000 |
| 1970 | ↓ | 16,000,000 |
| 1971 | 4,190,000 | 18,020,000 |

Source:  First National Bank of New York.

Evidently, U.S. reliance on much-enlarged quantities of oil imports will mean a considerable shift of such purchases from the Western Hemisphere to the Eastern Hemisphere. In spite of increased activity by the international companies in new areas of oil discoveries in the Eastern Hemisphere, the bulk of oil production still concentrates heavily in the several Arab oil-exporting countries. Thus, it is inevitable that the future association between these few Arab nations, the oil companies who now act as their agents, and the consuming nations, including the United States, would weigh heavily on the U.S. scale of policy priorities.

In weighing U.S. economic options in relation to Arab oil negotiations, the position of the international companies in the Middle East becomes all-important. Since the age of concessions has all but ended, these companies are no longer the real owners of their products. More often, the companies appear to be employed as these governments' business agents. Prior to the October 1973 war in the Middle East the companies' arrangements with the oil-exporting nations fell into two categories: participation and nationalization. It was widely believed, and with good reason, that all participation contracts were of very short duration and that, eventually, sooner or later, participation would lead to full nationalization.

Participation contracts were started with four Middle East countries early in 1973 with Saudi Arabia taking the lead.* On January 1, these states acquired 25 percent ownership in the international companies' producing operations. This ratio was to grow gradually until it reached 51 percent in 1982, at which time the companies would no longer have a deciding vote in their operations. This plan came to an abrupt end as a result of the outbreak of war in the Middle East in October 1973, when the participation contracts were canceled and all oil production in these countries was nationalized.

The participation agreements, however, did not include any part of the companies' transportation or distributive organizations. Rather than buy into these foreign operations, the Arab countries seemed to prefer to combine their efforts within the framework of OAPEC, in order to build their own capacities. According to this plan, by 1982, when, according to the parallel participation agreements,

_____

*Known as the Yamani Group, after the Saudi Minister of Oil who conducted the participation negotiations for Saudi Arabia, the group also included Abu Dhabi, Kuwait, and Qatar.

they were to become majority owners of the oil-producing operations in their countries, these nations will not have to rely to a very large extent on Western agents for the transportation and distribution of their oil. Considering that after the nationalization of the Anglo-Iranian Oil Company (AIOC) the private companies were able to boycott all Iranian production in Western markets by the simple fact of owning all transportation and distribution facilities, it was quite easy to understand why the Arab nations, prior to complete nationalization of all their oil resources, preferred to build their own facilities.

The sudden nationalization of these countries' oil resources and production was done in sharp contradiction to their former cautious policy that strove to avoid any possible Western threat of boycott. That the fear of a possible Western boycott of the oil of individual states as a retaliatory form for nationalization of companies' assets was very much alive and real prior to the October War can be further seen in the case of the nationalization of the IPC by the Iraqi government on June 1, 1972, while, at the same time, this government was closing an expanded oil sales deal with the USSR in exchange for Soviet technical and financial aid in developing the North Rumalia oil fields, a concession formerly held by the British company. Fearing a retaliatory boycott by the Western international concerns against Iraq, the Eighth Arab Petroleum Congress, which was holding its meeting at that time, "called on all Arab oil exporting countries to support Iraq in its nationalization program," by boycotting "any companies trying to block the Iraqi oil sales efforts."[7]

Owning ancillary transportation and distribution operations will not only increase the OAPEC members' capacity to unilaterally dictate posted prices and interfere with the free flow of oil as they please but will also release them from certain external bargaining pressures that are part and parcel of any free market negotiating practice. In short, what OAPEC is after is a worldwide monopoly of refining, transportation, and distribution operations of all oil products originating with Arab crude. Such a monopoly was never enjoyed by the international oil cartel even in its heyday.

Furthermore, the greater power the Arab nations gain in fixing the world oil market, the more hard currency they will accumulate in their exchequers. The extent to which these accumulated fortunes can influence the international monetary system and thereby establish new political

alignments depends on whether they will remain passive accounts or be spent.  No one argues that passive accounts is
in the consuming nations' best interest.  On the contrary, the
Western nations need vigorously circulating currencies in
order to keep their economies growing.  The question, then,
is to what end will the Arab nations spend or invest their
money:  Will this expenditure be aimed at nation-building
in their own countries or will much of the money be spent in
order to gain political leverage over other nations?

It is exactly this great concentration of economic, political, and strategic power in the hands of a few semifeudal
and largely unfriendly nations as well as nations who are
under tutelage of a rival great power that makes any increased U.S. dependence on Arab oil imports a problematical
matter at best.  The United States must not play these nations' game, for an unrestricted dependence upon these oil
imports may result in an increasing deficit in U.S. balance
of payments and a possible blackmailing of a good part of
U.S. foreign policy.  The current shortage of readily available oil in the United States, due to insufficient exploration and conservation of the domestic resources, calls for
expanded efforts in all areas of domestic energy.

## DOLLARS FOR SUPERTANKERS, OFFSHORE
## TERMINALS, AND INFLATION

Senator Henry M. Jackson, chairman of the Senate Committee on the Interior, who has held congressional hearings
on matters of U.S. energy resources, has offered the nation
a solution by way of a 10-year crash program of research and
development of all domestic energy resources in order to reduce drastically, or even eliminate completely, the need for
dependence on foreign imports.  The price tag of this crash
program, according to Jackson, would be $20 billion.  In
December 1973 Congress passed Jackson's energy bill and
granted the $20 billion.  This considerable amount of money
inhibits many of those who would like to see major cutdowns
made in government spending.  Some of these critics would
prefer to import Arab oil indefinitely rather than "reck-
lessly squander" this kind of money.

But what kind of expenditure will, in effect, accrue as
a result of an unrestricted U.S. dependence on Arab oil?  By
1985, it is estimated, Arab oil imports may be adding at
least $20 billion and quite possibly as much as $30 billion
(depending on the escalation of both prices and demands) to

235

the debit side of the U.S. balance of payments.* The cost
of Arab crude will not be cheaper than Jackson's program,
and the money paid to the Arab nations will be added to the
general dollar drain, which, in turn, will augment the rapid
rate of downhill inflation.

However, the cost of Arab oil is not the only expenditure that the United States will be forced to make. In
order to import crude oil, the United States must first have
at its disposal the means of transportation for the oil's
conveyance. Even the most ardent supporters of extensive
and indefinite dependence on Arab oil imports would consider
a heavy reliance on non-American-flag tankers a foolhardy
gamble with U.S. national security. It follows that in
order to import oil, the United States must have its own
tanker line. Such a fleet is not at present available, nor
can it possibly be constructed in time for the big rush on
Middle East oil. To carry U.S. Middle East imports only
seven years from 1973, the United States will have to build
a fleet the equivalent of 350 new supertankers of 250,000
deadweight tons each, at a conservatively estimated total
cost of $16-18 billion.† In order to qualify for government
subsidies, these vessels will have to be registered under
the U.S. flag for a period of only 10 years, at which time
their owners may find good reasons to transfer their registration to foreign flags of convenience. Thus, within 10
years, the United States stands to spend close to $20 bil-
lion on a fleet a large part of which it is bound to lose at
the end of that period.

Moreover, the pace of construction in the United States,
because of the small number of shipyards equipped to build
supertankers and because of the scarcity of trained labor to
do it, will lag far behind the escalating demands for oil.
Under the present laws, government subsidies for tanker construction are granted only for such vessels as are ordered
from U.S. shipyards. That means that unlike other nations'
the U.S. construction rate will never catch up with demand,
unless, in order to alleviate the situation, the nation will

---

*These figures are based on estimates made prior to the
recent price hikes of Middle East oil. No new estimates are
yet available as this book is going to print.

†Supertankers on order as of 1973 or under construction in foreign shipyards are, generally, in the 350,000-
deadweight-ton classification, with very few planned at
present (until more larger and deeper-water terminals are
available) of the 500,000-600,000 deadweight ton class.

again invest several billion dollars in building new shipyards, the construction of which alone, even before the
first tanker is built there, will take several years and
add more billions of dollars to oil expenditure.  On the
other hand, if the law is modified to allow the ship companies to place orders for new vessels in foreign shipyards,
the country will lose at least $15 billion, thereby burden-
ing even more its balance-of-payments program.  And, at the
same time, U.S. craftsmen will be deprived of rightful work
by the use of foreign labor.

If this staggering tanker expenditure is not enough,
other government spending will mount up to make Arab oil imports prohibitive.  Most important of all, because no vessel
larger than 70,000 tons can sail into any East Coast port
under a full load, it will be necessary to build special
offshore terminals complete with refilling storage, repair
shops, and pumping facilities for unloading the imported oil
from the ocean-going vessel either into a large number of
small coastwise tankers or into underwater pipelines, which
will convey the oil from the offshore terminal to the coastal
refineries.  The minimum cost of such an offshore terminal
island is $500 million.  Add the cost of this program to the
cost of the proposed supertanker fleet and, before even adding the cost of actual Arab oil imports, we arrive at a figure that with the escalating cost of maintenance and materials, will easily exceed $50 billion.

So far Congress has appropriated only $842 million for
construction subsidies of all seagoing vessels.  Only 46
percent of this amount, about $387,580,000, was allocated
by the Maritime Administration for the construction of tankers and liquefied natural gas carriers.  Even if Congress
approves an additional $275 million for this program for the
fiscal year 1973/1974, the total thus far will not add up to
more than $662,580,000, which is not even $1 billion.  At
this rate, by 1980 Congress will not have appropriated for
this program even $2 billion, which is less than 4 percent
of the total amount required in order to shift U.S. energy
dependence to Middle East oil imports.

Whether U.S. tankers are built in the United States or
in foreign shipyards, whether oil imports are conveyed
ashore via underwater pipelines or coastal tanker fleets,
this spending will have to be recovered out of the price of
oil.  By the time this oil reaches the United States, its
price will already be inflated due to increasing demands by
the producing nations, which will, at that time, hold the
monopoly over all production operations.  Although the U.S.
government's intention has been to confine foreign oil

purchases to crude in order to save dollars and ease the balance-of-payments burden, it is quite conceivable that OAPEC will demand that a considerable ratio of oil orders be made directly from local Arab refineries, in which case the United States will find itself paying even more for finished products. Thus, with the increase of the price of oil at the well-head and the increase of the price of oil due to the enormous construction cost as well as the generally escalating cost of maintenance and running of any operation, it is estimated that the American consumer will have to pay eight times more for his oil in 1985 than he is paying in 1973.

Of course, another realistic possibility with which the United States may be faced after it completes its construction program is that OAPEC will endorse a resolution by which all Arab crude and products must be carried in Arab bottoms. If that happens, oil expenses will be once again increased to the detriment of balance of payments; it will add to consumer's cost of oil; and it will endanger national security. When that occurs, the citizens of the United States will be enjoying the rare sight of a gigantic fleet of brand new and ultramodern supertankers, built at a very high cost, sitting idly until it is ready to be collected for scrap.

## SUMMARY AND CONCLUSIONS

Any increased dependence of the U.S. economy on Middle East oil and, as a consequence, on the future reliability of a small group of Arab nations and Iran must be appreciated against the backdrop of regional realities as well as U.S. worldwide political and strategic priorities.

Since the first Soviet arms shipments to Egypt in 1955, it has been a cornerstone of U.S. foreign policy in the Middle East to isolate commercial oil considerations from the political and military imperatives inherent in the geopolitical complex of this region. As a result, international oil companies having operations in the Middle East soon found that they no longer could depend on State Department action on their behalf when their contracts or assets were being violated by the host governments. For nearly 20 years, U.S. diplomacy has been pursuing, first, Arab regional alliances and, second, Arab friendship, in open preference to its traditional role of protecting the property, property rights, and commercial opportunity of U.S. citizens abroad.

The single factor that seems to have motivated this deviation from long-standing principles of U.S. foreign

policy is the apparent increase of Soviet influence in many
countries of the Middle East and the Indian Ocean and the
concurrent decrease in U.S. influence.  It will appear, then,
that there has never been a greater need for U.S. Middle
East policy to sacrifice commercial considerations for the
security of political and military interests.  If that security eludes the United States, the commercial interests will
be completely at the mercy of the host governments.  If, on
the other hand, the United States strives to maintain Middle
East equilibrium in its favor, the U.S. commercial corporation doing business in the Middle East will also find the
area's general climate better accommodating its interests.
To achieve this, the United States must free itself from
economic dependence on uncertain sources.  It must rapidly
develop its own domestic natural and manufactured energy
resources, until it becomes once again reasonably self-
sufficient.  An independent energy economy will bolster
national security and broaden foreign policy options.

# NOTES

CHAPTER 1

1.   U.S. Congress, Senate, Select Committee on Small Business, _The International Petroleum Cartel_, a staff Report of the Federal Trade Commission submitted to the Subcommittee on Monopoly (Washington, D.C.:   Government Printing Office, August 22, 1952), p. 45.

2.   Blair Bolles, "Oil:   An Economic Key to Peace," _Foreign Policy Reports_ 20, 8 (July 1, 1944):   91-92.

3.   J. H. Carmical, _The Anglo-American Petroleum Pact_ (New York and Washington, D.C.:   American Enterprises Association, 1945), p. 44.

4.   Herbert Feis, _Petroleum and American Foreign Policy_, Commodities Policy Study No. 3, Food Research Institute of Stanford University (Stanford, Cal., March 1944), p. 3.

5.   _Congressional Record_, 67th Cong., 1st sess., April 12, 1921, pp. 166-167.

6.   Feis, op. cit., p. 5.

7.   U.S. Department of State, _Oil Concessions in Foreign Countries_, Doct. 97, 68th Cong., 1st sess.   (Washington, D.C.:   Government Printing Office, 1924).

8.   Feis, op. cit., p. 6.

9.   U.S. Congress, Senate Doct. 11, 67th Cong., 1st sess., p. 25.

10.   U.S. Department of State, _Papers Relating to the Foreign Relations of the United States_, 1936 (Washington, D.C.:   Government Printing Office, 1942), vol. II, p. 664.

11.   Ibid., 1927, vol. II, pp. 813, 824.

12.   Herbert Feis, _The Diplomacy of the Dollar:   First Era, 1919-1932_ (Baltimore:   Johns Hopkins Press; Hampden, Conn.:   Archon Book, 1965), p. 58.

13.   Christopher Tugendhat, _Oil:   The Biggest Business_ (New York:   G. P. Putnam's Sons, 1968), p. 89.

14.   U.S. Congress, Senate Special Committee Investigating Petroleum Resources, _American Petroleum Investments in Foreign Countries_, 79th Cong., 1st sess. (1946), p. 32.   Politically, the risks encountered by foreign oil operations could range from "governmental competition, compulsory refinery installation, and trade and exchange control on the one end, to prevention or cancellation of contracts and outright expropriation on the other" (p. 40).

15. New York _Times_, March 29, 1933, p. 27.

16. _Ibid._, February 9, 1933, p. 25. According to George W. Stocking and Myron W. Watkins, "557 basic codes . . . blanketed American industry and . . . 'stabilized' business by subordinating the independents to the vested interests of large corporations" (_Monopoly and Free Enterprise_ [New York: Greenwood Press, 1968], p. 44).

17. _Ibid._, June 11, 1933, II, p. 1; June 18, 1933, p. 9; June 25, 1933, II, p. 13; August 19, 1933, p. 2.

18. Temporary National Economic Committee, _Investigation of the Concentration of Economic Power_, Hearings pursuant to Pun. Res. 113 (75th Congress), cited in Stocking and Watkins, op. cit., p. 44.

19. National Petroleum Code, Text, New York _Times_, July 15, 1933.

20. _Ibid._, August 30, 1933.

21. R. C. Holmes to Franklin Delano Roosevelt, a letter, New York _Times_, April 3, 1933, p. 26; also, see ibid., June 16, 1933, p. 17 and July 24, 1933, p. 27.

22. Raymond A. Hare, "The Great Divide: World War II," _Annals_ of the American Academy of Political and Social Science, 401 (May 1972): 24.

23. Cited in Feis, _Diplomacy of the Dollar_, op. cit., p. 50.

24. _Ibid._, p. 48.

CHAPTER 2

1. Harold L. Ickes, _Fighting Oil_ (New York: Alfred A. Knopf, 1943), p. 23.

2. Charles W. Hamilton, _Americans and Oil in the Middle East_ (Houston, Texas: Gulf Publishers Company, 1962), p. 150.

3. U.S. Department of Defense, Munitions Board Petroleum Committee, "Foreign Oil" (Washington, D.C., mimeographed, December 1949, formerly "restricted" material), vol. I: "The Middle East and Africa," p. 1.

4. _Economist_ 146, 5242 (February 12, 1944): 213.

5. Ickes, op. cit., p. 142.

6. Herbert Feis, _Petroleum and American Foreign Policy_ (Stanford, Cal., March 1944), pp. 15-16.

7. Benjamin Shwadran, _The Middle East, Oil and the Great Powers_ (New York: Praeger, 1956), p. 304.

8. U.S. Congress, Senate, Special Committee Investigating the National Defense Program, _Petroleum Agreements with Saudi Arabia_, Hearings, Part 41, 80th Cong., 1st sess. (Washington, D.C.: Government Printing Office, 1948), pp. 24725-24726.

9.  Ibid., p. 24801.
10.  Ibid.
11.  Herbert Feis, The Diplomacy of the Dollar: First Era, 1919-1932 (Baltimore:  Johns Hopkins Press, 1950; Hampden, Conn.:  Archon Book, 1965), p. 60.
12.  Senate Special Committee Investigating National Defense Program, op. cit., p. 311 and Hamilton, op. cit., pp. 150-151.
13.  New York Times, October 8, 1943, p. 8.
14.  Petroleum Reserves Corporation, Technical Oil Mission to the Middle East, Preliminary Report (Washington, D.C.:  Government Printing Office, February 1, 1944); also printed in Congressional Record, Senate, 94, part 4, 4962, April 28, 1948.
15.  Hamilton, op. cit., p. 151.
16.  Shwadran, op. cit., pp. 318-319.
17.  New York Times, November 18, 1943, p. 38.
18.  Feis, Petroleum, op. cit., p. 44.
19.  Cited in Economist 146, 5242 (February 12, 1944):  211.
20.  Feis, Petroleum, op. cit., p. 44.
21.  Herbert Feis, Seen From E.A. (New York:  Alfred A. Knopf, 1947), p. 140.
22.  New York Times, January 28, 1944, p. 1.
23.  Feis, Petroleum, op. cit., p. 2.
24.  Economist 146, 5243 (February 19, 1944):  240.
25.  Feis, E.A., op. cit., p. 152.
26.  Ibid.
27.  Ibid.
28.  New York Times, March 3, 1944.
29.  Ibid., March 7, 1944, p. 11.
30.  Ibid.
31.  Ibid., March 5, 1944, pt. VI, p. 5.
32.  Ibid.
33.  Ibid.
34.  Feis, E.A., op. cit., p. 154.

CHAPTER 3

1.  New York Times, March 30, 1944, p. 1.
2.  George Lenczowski, Russia and the West in Iran, 1918-1948 (New York:  Greenwood Press, c. 1968), p. 216.
3.  New York Times, October 11, 1944, p. 6.
4.  Ibid., October 18, 1944, p. 10.
5.  Ibid., October 29, 1944, pt. III, p. 6.
6.  Ibid., December 4, 1944, p. 11.
7.  Ibid., December 21, 1944, p. 4.

8.  Ibid., October 29, 1944, pt. III, p. 6.  As against this figure, Wallace Pratt, vice president of Standard Oil Company of New Jersey, asserted that "if we look at the petroleum reserves of the first rank nations, we find Russia claiming first place.  In the years imme- diately preceding the war Russia published estimates which placed her proved reserves as 45 billion barrels." Adding that Russia's potential reserves were indeed "enormous," he estimated the current world's proved re- serves at more than 100 billion barrels, of which the United States possessed 20 billion and the USSR owned 45 billion.  Wallace Pratt, "Undiscovered Reserves Equal to Those Already Found Held Probable," Oil and Gas Journal 42, 34 (December 30, 1943):  78.

9.  "The Iranian Oil Question," New Republic 111, 20 (November 13, 1944):  615.

10.  New York Times, December 21, 1944, p. 7.

11.  Ibid., March 17, 1945, p. 17.

12.  Ibid.

13.  Ibid., October 29, 1944, pt. IV, p. 4.

14.  "Russia and the Middle East," Economist 147, 5278 (October 21, 1944):  532.

15.  Ibid.

16.  New York Times, November 28, 1945, p. 26.

17.  New Republic 111, 20 (November 13, 1944):  615.

18.  Herbert Feis, Seen from E.A. (New York:  Alfred A. Knopf, 1947), p. 156.

19.  New York Times, March 8, 1945, p. 7.

20.  For Brewster's views see New York Times, Sep- tember 30, 1943, p. 7; for Lodge's views, see ibid., October 13, 1943, p. 9.  The quotation is by Brewster.

21.  James Francis Byrnes, "U.S. Aims and Policies in Europe," an address delivered in Paris on October 3, 1940, at the American Club, Department of State Publica- tion 1670, European Series 18 (Washington, D.C.:  Govern- ment Printing Office, 1946), p. 3.

22.  New York Times, December 15, 1943, p. 37.

23.  Ibid.

24.  Ibid.

25.  Ibid.

26.  Feis, E.A., op. cit., p. 135.

27.  Ibid., p. 134.

28.  Ibid., p. 135.

29.  New York Times, February 11, 1944, p. 1.

30.  Ibid.

31.  Ibid., February 12, 1944, p. 28.

32.  Ibid., p. 1.

33. Ibid., February 11, 1944, p. 1.

34. _Economist_ 146, 5251 (April 15, 1944): 509.

35. The text of the Atlantic Charter was published in U.S. Congress, House of Representatives, _Message of the President_, House Document No. 358, 77th Cong., 1st sess. (Washington, D.C.: Government Printing Office, 1941); and in U.S. Department of State, _Foreign Relations of the United States: Diplomatic Papers 1941_ (Washington, D.C.: Government Printing Office, 1963), vol. I, pp. 341-78: "Joint Statement by President Roosevelt and Prime Minister Churchill, August 14, 1941," ibid., pp. 367-68.

36. George Lenczowski, _Oil and State in the Middle East_ (Ithaca, N.Y.: Cornell University Press, 1960), p. 169.

37. Great Britain, Treaties, _Agreement on Petroleum_, United States No. 1, Cmd. 6555 (London: His Majesty's Stationery Office, 1944), 6 pages.

38. U.S. Congress, Senate Special Committee Investigating Petroleum Resources, _American Petroleum Investments in Foreign Countries_ (hereafter, APIIFC Hearings), 1945, pp. 8-9.

39. J. H. Carmical, _The Anglo-American Petroleum Pact_, economic survey series no. 408 (New York: American Enterprise Association, 1945), pp. 18-22.

40. Feis, _E.A._, op. cit., p. 184.

41. Ibid.

42. _Economist_ 146, 5251 (April 15, 1944): 509.

43. Ibid., 147, 5268 (August 12, 1944): 225.

44. Carmical, op. cit., p. 7.

45. Feis, _E.A._, op. cit., p. 166 (emphasis added).

46. Early in 1944, President Roosevelt appointed the following petroleum committee to handle the negotiations with the British: Cordell Hull, secretary of State, chairman; Harold L. Ickes, secretary of the Interior and Petroleum Administrator for War, vice chairman; James Forrestal, undersecretary of the Navy (at that time); Robert P. Patterson, undersecretary of War; Leo T. Crowley, Foreign Economic Production Board; Ralph K. Davis, deputy petroleum administrator for War; Charles Rayner, petroleum adviser, Department of State. U.S. Congress, Senate, Committee on Foreign Relations, _Petroleum Agreement with Great Britain and Northern Ireland_, Hearings, June 1947, 80th Cong., 1st sess. (Washington, D.C.: Government Printing Office, 1947), statement of Charles Rayner, petroleum adviser, Department of State, p. 31.

47.  Michael Brooks, <u>Oil and Foreign Policy</u> (London: Lawrence and Wishart, 1949), p. 97.
48.  Economist 147, 5270 (August 26, 1944):  273.
49.  Feis, <u>E.A.</u>, op. cit., p. 169.
50.  The text of the revised Anglo-American Petro-leum Agreement was published in full in the following government documents:  U.S. Congress, Senate, Committee on Foreign Relations, <u>Anglo-American Oil Agreement</u>, Re-port to accompany Executive H, 79th Cong., lst sess., July 1, 1947 (Washington, D.C.:  Government Printing Of-fice, 1947); and <u>Petroleum Agreement with Great Britain and North Ireland</u>, Hearings, 80th Cong., lst sess. (Washington, D.C.:  Government Printing Office, 1947), pp. 3-5.
51.  Brooks, op. cit., p. 132.
52.  Lenczowski, op. cit., p. 171.
53.  New York <u>Times</u>, December 15, 1943, p. 37.
54.  Brodie, op. cit., p. 298.
55.  Ibid., p. 301.
56.  Ibid.
57.  Feis, <u>Petroleum</u>, op. cit., p. 17.
58.  Ibid., p. 18.
59.  Ibid., p. 29.
60.  Senate Special Committee, <u>APIIFC Hearings</u>, op. cit., pp. 27-28.
61.  Ibid., p. 45.
62.  Ibid., p. 46.
63.  Ibid., p. 39.
64.  Ibid.
65.  Ibid., p. 38.
66.  Ibid., p. 46, statement of Brig. Gen. Peckham, Liaison Officer for Petroleum, War Department.
67.  J. C. Hurewitz, <u>Middle East Politics, The Mili-tary Dimension</u> (New York:  Praeger Publishers, 1969), pp. 70-72.
68.  Senate Special Committee Investigating Petro-leum Resources, <u>Investigation of Petroleum Resources in Relation to the National Welfare</u>, <u>Final Report</u>, Report no. 9, 80th Cong., lst sess. (Washington, D.C., 1947), pp. 27-28.
69.  Ibid., p. 54.
70.  Ibid., p. 8.
71.  Ibid., pp. 45-46.
72.  Ibid.
73.  New York <u>Times</u>, January 7, 1944, pp. 7 and 8.
74.  United Nations Economic and Social Council, <u>Control of Oil Resources</u>, item proposed by the

International Cooperative Alliance, E/449, July 2, 1947, p. 2. (mimeo.)

75. Ibid.

76. Ibid., p. 3.

77. United Nations Economic and Social Council, Control of World Oil Resources, item 26, a proposal for the creation of a United Nations Petroleum Commission under the authority of the Economic and Social Council, presented by the International Cooperative Alliance, supporting document no. 2, E/449/Add.1, July 31, 1947, p. 2. (mimeo.)

78. United Nations Economic and Social Council, Doct. E/449, op. cit., p. 3.

79. United States Reports, United States vs. Socony-Vacuum Co. Inc. et al., argued February 5, 6, 1940 --decided May 6, 1940, Cases Adjudged in the Supreme Court at October Term, 1939, volume 310 (Washington, D.C.: Government Printing Office, 1940), pp. 221-223.

80. Ibid.

81. United States Statutes at Large, "An Act to Promote Export Trade, and for Other Purposes," H.R. 2316, April 10, 1918, volume 40 (Washington, D.C.: Government Printing Office, 1919), pp. 516-518.

82. National Petroleum Council, A National Petroleum Policy for the United States (Washington, D.C., 1949), p. 23.

83. Ibid., p. 3.

84. Ibid., p. 5.

85. Ibid.

86. National Petroleum Council, Committee on National Petroleum Emergency, "Report" (mimeographed for interoffice use, January 13, 1949), pp. 1-2.

87. Ibid., p. 2.

88. Ibid., p. 3.

89. Defense Production Act (1950), Public Law 775, Section 708(a).

CHAPTER 4

1. U.S. Congress, Senate, Committee on Interstate and Foreign Commerce, Controlling the Movement and Transportation Abroad of Gasoline and Petroleum Products, 80th Cong., 1st sess., Report No. 696 (Washington, D.C.: Government Printing Office, 1947).

2. Ibid., p. 6. For further discussion of this piece of legislation by the House of Representatives, see U.S. Congress, House of Representatives, Committee on Merchant Marine and Fisheries, Controlling Movement and

<u>Transportation of Gasoline and Petroleum Products</u>, Report to accompany H.R. 4042, July 21, 1947, 80th Cong. (Washington, D.C.: Government Printing Office, 1947), 8 pp.

3. U.S. Department of the Interior, <u>National Resources and Foreign Aid</u>, Report of J. A. Krug, Secretary of the Interior, October 9, 1947 (Washington, D.C.: Government Printing Office, 1947), 97 pp.

4. United States President, Office of Government Reports, <u>European Recovery and American Aid</u>, Report by President's Committee on Foreign Aid, November 7, 1947 (Washington, D.C.: Government Printing Office, 1947), 286 pp.

5. U.S. Congress, House of Representatives, Select Committee on Foreign Aid, <u>Petroleum Requirements and Availabilities</u>, Preliminary Report No. 5, 80th Cong., 1st sess. (Washington, D.C.: Government Printing Office, 1947).

6. U.S. Congress, Senate, Committee on Foreign Relations, <u>Outline of European Recovery Program</u>, draft legislation and background information submitted by the Department of State for use of Senate Foreign Relations Committee, December 19, 1947 (Washington, D.C.: Government Printing Office, 1948), 131 pp.

7. U.S. Congress, House of Representatives, House Reports, <u>Foreign Assistance Act of 1948</u>, conference report to accompany S. 2202, April 1, 1948 (Washington, D.C.: Government Printing Office, 1948), 35 pp. See also, U.S. Public Laws, <u>Act to Promote World Peace and General Welfare, National Interest, and Foreign Policy of United States through Economic, Financial, and other measures necessary to maintenance of conditions abroad in which Free Institutions may survive and consistent with maintenance of strength and stability of United States</u>, approved April 3, 1948, 80th Cong., 2d sess. (Washington, D.C.: Government Printing Office, 1948), 26 pp.

8. U.S. Department of State, <u>Committee of European Cooperation</u>, vol. 1: General Report, vol. 2: Technical Reports (Washington, D.C.: Government Printing Office, 1947).

9. U.S. Department of State, Public Affairs Office, Publications Division, <u>European Recovery Program: Commodity Reports Including Manpower</u>, Economic Cooperation Series, Publication No. 3093 (Washington, D.C.: Government Printing Office, 1948), 448 pp.

10. Elmer Patman, <u>The Third World Petroleum Congress</u>, a report to the Select Committee on Small Business, U.S. Senate, and the Select Committee on Small Business of the

House of Representatives, 82d Cong., 2d sess. (Washington, D.C.: Government Printing Office, 1952), p. 45.

11. Organization for European Economic Cooperation (OEEC), Oil Committee, First Report on Co-ordination of Oil Refinery Expansion in the O.E.E.C. Countries, Report by the Oil Committee on the Long-Term Plans of the Participating Countries (Paris, October 1949), pp. 5-6.

12. OEEC, Third Report on Co-ordination of Oil Refinery Expansion in the O.E.E.C. Countries (Paris, June 1953), p. 7.

13. Ibid.

14. Ibid., p. 21.

15. Ibid., p. 25.

16. The British note to the French government in regard to the proposed Locarno Treaties "expressed a hope that the United States might 'find it possible to associate themselves with the agreements which would thus be realized.'" Great Britain, Treaties, Franco-British Correspondence, White Paper, Cmd. 2435, cited in Survey of International Affairs 1925 (London: Oxford University Press, 1928), vol. 2, p. 39.

On the occasion of his speaking at Cambridge, Massachusetts, on July 2, 1925, Coolidge addressed himself to this note, cautioning "that while the United States should avoid political commitments where it has no political interests, such covenants would always have our Government's moral support. . . ." (cited ibid., p. 62). The following October, when the terms of the Locarno Pact became known in Washington, Coolidge again expressed "deep gratification that Europe had shown readiness to help itself; the United States, he declared, was ready to do all in its power to give aid, short of jeopardizing its own interests . . ." (ibid.).

17. James Francis Byrnes, Moscow Meeting of Foreign Ministers, December 16-25, 1945, and Soviet-Anglo-American Communiqué, Department of State Publication 2448, Conference Series 79 (Washington, D.C.: Government Printing Office, 1946), p. 3.

18. Ibid.

19. Ibid., p. 8.

20. Arnold Wolfers, ed., Alliance Policy in the Cold War (Baltimore: Johns Hopkins Press, 1959), p. 1.

21. Alastair Buchan, Europe's Futures, Europe's Choices (New York: Columbia University Press for the London Institute of Strategic Studies, 1969), p. 4.

CHAPTER 5

1. New York *Times*, January 30, 1951, p. 24.
2. Ibid., January 3, 1951, p. 82. Total U.S. aid
to Iran up to 1951 was under $50 million, in spite of
many promises of more. See letter to the editor, Helen
Kennedy Stevens, ibid., February 22, 1951, p. 6.
3. Ibid., January 16, 1951, p. 6.
4. Ibid., January 14, 1951, p. 12.
5. Ibid., January 9, 1951.
6. Ibid., March 2, 1951, p. 5.
7. Ibid., March 9, 1951, pp. 1 and 5.
8. Ibid., March 19, 1951, pp. 1 and 5.
9. Ibid., March 28, 1951, p. 6.
10. George Lenczowski, *The Middle East in World
Affairs* (Ithaca, N.Y.: Cornell University Press, 1962),
p. 206.
11. U.S. Department of the Interior, Office of Oil
and Gas, *The Middle East Petroleum Emergency of 1967*
(Washington, D.C.: Government Printing Office, 1969),
Volume II, p. A-1.
12. Ibid.
13. U.S. Defense Production Administration, Defense
Production Program, *Defense Production Record*, vol. I,
no. 11 (July 12, 1951), p. 6, and vol. I, no. 17 (August
23, 1951), p. 6.
14. Ibid., p. 186.
15. OEEC, Oil Committee, *Third Report on Oil Refinery Expansion in the OEEC Countries* (Paris, June 1953),
p. 16.
16. Ibid., p. 17.
17. Ibid., p. 25.
18. Ibid., p. 18.
19. Harry N. Howard, "The Development of United
States Policy in the Near East, South Asia and Africa
during 1954," U.S. Department of State *Bulletin* 32, 816
(February 14, 1955): 257.
20. George W. Stocking, *Middle East Oil: A Study
in Political and Economic Controversy* (Kingsport, Tenn.:
Vanderbilt University Press, 1970), pp. 158-159.
21. Ibid., p. 157.
22. Ibid.
23. Ibid.
24. Emanuel Celler, "Concentration of Economic
Power Can Lead to Government Control and Inefficiency,"
*Vital Speeches* 21, 17 (June 15, 1956): 529.

25.  Stocking, op. cit., p. 158.

26.  Assistant Secretary of Defense for International Security Affairs, H. Struve Hensel, "Effect of the Cold War on Foreign Economic Policy," Vital Speeches 21, 11 (March 15, 1955): 1099.

27.  George Lenczowski, "United States Support to Iranian Independence," Annals of the American Academy of Political and Social Science 401 (May 1972): 50.


CHAPTER 6


1.  U.S. Congress, House of Representatives, Committee on Merchant Marine and Fisheries, The Merchant Marine Act of 1936 and the Ship Sales Act of 1946, Hearings, 80th Cong., 2d sess. (Washington, D.C.: Government Printing Office, 1948), p. 82.

2.  Ibid.

3.  Ibid., p. 533.

4.  U.S. Department of the Interior, Petroleum Administration for Defense, Supply and Transportation Division, Transportation of Oil (Washington, D.C.: Government Printing Office, December 1951), p. 8.

5.  U.S. Congress, Senate, Committee on Interstate and Insular Affairs, Cargo Preference Bill, Hearings Before a Subcommittee on Water Transportation, 83d Cong., 2d sess. (Washington, D.C.: Government Printing Office, 1954), letter from the Department of the Navy, pp. 5-6.

6.  Organization for European Economic Cooperation (OEEC), Maritime Transport (Paris, 1954), p. 63.

7.  U.S. Congress, Senate, Committee on Interstate and Foreign Commerce, Cargo Preference Bill, Hearings Before a Subcommittee on Water Transportation, 83d Cong., 2d sess. (Washington, D.C.: Government Printing Office, 1954), statement of Arthur G. Syran, Director of Transportation, Foreign Operations Administration, p. 42.

8.  OEEC, op. cit., p. 5.

9.  Lloyd's Register Shipbuilding Returns, June 1954.

10.  Organization for European Economic Cooperation, Maritime Transport (Paris, 1955), p. 16.

11.  Organization for European Economic Cooperation, Maritime Transport (Paris, 1957), p. 61.

12.  Ibid., p. 22.

13.  OEEC, Maritime Transport, 1955, op. cit., p. 63.

14.  Ibid., p. 64.

15.  OEEC, Maritime Transport, 1954, op. cit., p. 11.

16.  Ibid., p. 13.

17.  Organization for European Economic Cooperation, Europe's Growing Needs of Energy (Paris, 1956), p. 13. This publication is popularly referred to as the Hartley report, thus named after the Hartley Commission, which authored it.

18.  Ibid., pp. 52-56.

19.  Organization for European Economic Cooperation, The Chemical Industry in Europe (Paris, 1953).

20.  Organization for European Economic Cooperation, The Chemical Industry in Europe (Paris, 1956), p. 91.


CHAPTER 7

1.  Manchester Guardian, The Record on Suez (Manchester:  Manchester Guardian Press, 1956), p. 3.

2.  Ibid.

3.  Ibid.

4.  The Suez Canal Company Nationalization Law was published in the following Egyptian government publications:  (1) Official Gazette, July 26, 1956, and (2) Republic of Egypt, Ministry of Foreign Affairs, White Paper on the Nationalization of the Suez Maritime Canal Company (Cairo:  Government Press, August 12, 1956), pp. 3-5.

5.  D. A. Farnie, East and West of Suez:  The Suez Canal in History 1854-1956 (Oxford:  Clarendon Press, 1969), p. 720.

6.  Sir Anthony Eden, Memoirs:  Full Circle (London: Cassell, 1960), p. 429.

7.  A. J. Barker, Suez:  The Seven Day War (New York:  Praeger Publishers, 1965), p. 20.

8.  Sir Anthony Eden, op. cit.

9.  Barker, op. cit.

10.  Ibid., p. 466.

11.  Ibid., p. 435.

12.  Ibid., p. 467.

13.  Robert Hunter, Security in Europe, International Relations Series, vol. 2 (London:  Elek Books, 1969), p. 111 (italics in text).

14.  John Marlowe, A History of Modern Egypt and Anglo-Egyptian Relations 1800-1956 (Hamden, Conn.: Archon Books, 1954; 2d ed., 1956), p. 367.

15.  Letter of Sir Anthony Eden to President Dwight D. Eisenhower, dated September 6, 1956, in Sir Anthony Eden, op. cit., p. 465.

16.  Ibid., p. 436.

17.  Ibid.

18.  Message from N. A. Bulganin to Sir Anthony Eden, in Soviet News, Suez and the Middle East:  Documents, a

second collection covering November 5-December 9, 1956 (London: Soviet News, 1956), p. 7.

19. U.S. Congress, Senate, Committee on the Judiciary and Committee on the Interior and Insular Affairs, Emergency Oil Lift Program and Related Oil Problems, Joint Hearings Before Subcommittees (EOLP Hearings), 85th Cong., 1st sess. (Washington, D.C.: Government Printing Office, 1957), testimony of W. A. Delaney, Ada, Oklahoma, p. 876.

20. Ibid., testimony of Edwin G. Moline, Officer in Charge, Economic Organization Section, Office of European Regional Affairs, Department of State, in answer to question by Senator Everett Dirksen, p. 600.

21. U.S. Department of the Interior, Office of Oil and Gas, Report to the Secretary of the Interior, from the Director of the Voluntary Agreement, as Amended May 8, 1956, Relating to Foreign Petroleum Supply (Washington, D.C.: mimeographed, June 30, 1957), p. 5.

22. Ibid., p. 5.

23. Ibid., p. 54.

24. Document presented under subpoena by Standard Oil Company of New Jersey, entitled, "Western Hemisphere Oil Supplies for Europe as of January 7, 1957,"--ibid., p. 2134.

25. Senate, EOLP Hearings, op. cit., p. 61.

26. Interior, Report, June 30, 1957, op. cit., p. 23.

27. Ibid., document presented under subpoena by New Jersey Standard, p. 2201.

28. Ibid., Statement of Russell B. Brown, General Counsel, Independent Petroleum Association of America, p. 371.

29. Ibid., Statement of Edwin G. Moline, p. 127.

30. Ibid., Statement of Fred A. Seaton, Secretary of the Interior, p. 572.

31. Ibid., Moline, p. 116.

32. Ibid., p. 61.

33. Erich W. Zimmermann, Conservation in the Production of Petroleum: A Study in Industrial Control, Petroleum Monograph Studies, vol. 2, American Petroleum Institute (New Haven, Conn.: Yale University Press, 1957), p. 130.

34. Ibid., pp. 190-191.

35. Ibid. As early as 1912 the navy, expounding a similar view, requested that the Department of the Interior set aside lands sufficient to ensure a supply of 500 million barrels of oil. In the course of time, four naval

reserves were set up:   reserves no. 1 and 2, Elk Hills
and Buena Vista Hills, in Karen County, California; reserve no. 3, Teapot Dome, in Wyoming; and reserve no. 4
in Alaska (p. 189).

36.   U.S. Congress, House of Representatives, Com-
mittee on Banking and Currency, Extension of the Defense
Production Act of 1950, Hearings, 85th Cong., 2d sess.
(Washington, D.C.:   Government Printing Office, 1951),
statement of Gordon Gray, director, Office of Defense
Mobilization, pp. 1-2.

37.   Ibid., introductory letter.

38.   Ibid.

39.   U.S. Congress, Senate, Committee on Banking
and Currency, Defense Production Act 1962, Hearings on
June 5 and 6, 1962, 87th Cong., 2d sess. (Washington,
D.C.:   Government Printing Office, 1962).   Statement of
Anthony A. Berston, Deputy Administrator, Business and
Defense Administration, U.S. Department of Commerce,
p. 19.

40.   Senate, EOLP Hearings, op. cit., statement of
Arthur S. Flemming, director, Office of Defense Mobilization, p. 20.

41.   Ibid., p. 563.

42.   Ibid., p. 487.

43.   U.S. Department of State Bulletin 35, 897
(September 3, 1956):   374.

44.   Senate, EOLP Hearings, op. cit., Senator
O'Mahoney, p. 1007.

45.   Ibid., statement of Coleman, p. 1017.

46.   Interior, Report, op. cit., p. 63.

47.   Ibid., pp. 16-18.

48.   Ibid., p. 4.

49.   Ibid., p. 10.

50.   Senate, EOLP Hearings, op. cit., p. 520.

51.   U.S. Congress, House of Representatives, Com-
mittee on Interstate and Foreign Commerce, Petroleum
Emergency, Hearings, 85th Cong., 1st sess. (Washington,
D.C.:   Government Printing Office, 1957), statement of
Hugh Stewart, pp. 29-30.

52.   Senate, EOLP Hearings, op. cit., pp. 1182-1185.

53.   Interior, Report, op. cit., p. 15.

54.   Ibid., p. 20.

55.   Senate, EOLP Hearings, op. cit., statement of
Felix E. Wormser, assistant secretary of the Interior,
p. 74.

56.   Interior, Report, op. cit., p. 31.

57.   Ibid., pp. 46-47.

58. Ibid., p. 57.

59. Ibid., p. 58.

60. Ibid., p. 34.

61. Ibid., pp. 59-60.

62. Senate, EOLP Hearings, op. cit., as stated by Senator Wiley, p. 8. Wiley was also a member of the Committee on Foreign Relations and for some time served as its chairman.

63. Ibid., p. 7.

64. Ibid., p. 912.

65. Ibid., statement of Russell B. Brown, general counsel, Independent Petroleum Association of America, p. 334.

66. Ibid., p. 341.

67. Ibid., statement of Assistant Secretary of the Interior Felix E. Wormser, p. 92.

68. Ibid.

69. See, for example, United Nations documents A/3215-3250, A/3252-3255, A/3257-3265, and A/3271, containing letters and cables from 50 Afro-Asian, Latin American, and communist nations. Also, see Resolution 999 (ES-I), Doct. A/3275, submitted by 19 Powers.

70. U.S. Department of State, Bulletin 38, 989 (June 9, 1958): 948.

71. Ibid. 35, 911 (December 10, 1956): 918.

72. U.S. Department of State, Bureau of Public Affairs, Historical Office, American Foreign Policy: Current Documents, Department of State Publication 7101 (Washington, D.C.: Government Printing Office, February 1961), documents relating to the enactment of the Middle East Resolution (the "Eisenhower Doctrine"), pp. 783-831; and "The Middle East Resolution," Public Law 7, House Joint Resolution 117, pp. 829-831.

73. U.S. Department of State, Bulletin 38, 989 (June 9, 1958): 948.

74. Royal Institute of International Affairs, Information Department, "The Western Powers and the Middle East 1959, A Documentary Record," A Chatham House Memorandum (London: mimeographed, distributed by Oxford University Press, June 1959), p. 11.

75. Hansard, Commons, vol. 64, col. 32, cited in ibid., p. 14.

76. U.S. Department of State, Bulletin 38, 989 (June 9, 1958).

77. Harold Lubell, Middle East Crises and Western Europe's Energy Supply (Baltimore: Johns Hopkins Press for RAND Corporation, 1963), p. xviii.

78. U.S. Department of State, Current Documents, Special Message of the President, p. 785 (emphasis added).

79. Ibid., p. 789.

80. Soviet News, Suez: The Soviet View, statements by the Soviet Government, N. A. Bulganin and D. T. Shepilov (London: Soviet News, 1956), pp. 3-5.

81. Ibid., p. 66.

82. Ibid., p. 67.

83. Ibid.

84. Paul Johnson, The Suez War, with a foreword by Aneurin Bevan (London: MacGibbon and Kee, 1957), p. 68.

CHAPTER 8

1. See, for example, a speech by Lenin on November 22, 1919, at the Second All-Russian Congress of Communist Organizations of the Peoples of the East.

2. Cited in Stanley W. Page, Lenin and World Revolution (New York: New York University Press, 1959; Gloucester, Mass.: Peter Smith, 1968), p. 155.

3. Collected Works, vol. 25, p. 290; cited in Joseph Stalin, Marxism and the National and Colonial Question, A Collection of Articles and Speeches (New York: Marxist-Leninist Library, International Publishers, 1935), p. 234.

4. Ibid., p. 233.

5. U.S. Congress, Senate, Committee on the Judiciary, Soviet Oil in the Cold War, A Study by Halford L. Hoskins and Leon Herman for the Library of Congress at the Request of the Subcommittee Investigating the Administration of the Internal Security Act and Other Internal Security Laws, 87th Cong., 1st sess. (Washington, D.C.: Government Printing Office, 1961), p. 1.

6. Frederick C. Barghoorn, The Soviet Cultural Offensive: The Role of Cultural Diplomacy in Soviet Foreign Policy (Princeton, N.J.: Princeton University Press, 1960), pp. 1 and 188.

7. Klaus Billerbrook, "Soviet Bloc Foreign Aid to the Underdeveloped Countries" (Hamburg: Hamburg Archives of World Economy, mimeographed, 1960), p. 16.

8. Under-Secretary Douglas Dillon, "Realities of Soviet Foreign Economic Policies," U.S. Department of State Bulletin 40, 1025 (February 16, 1959): 239.

9. U.S. Congress, House of Representatives, Committee on Foreign Affairs, The Soviet Economic Offensive in Western Europe, Report of the Special Study Mission

255

to Europe, House Report no. 32, 88th Cong., 1st sess. (Washington, D.C.: Government Printing Office, 1963), p. xi.

10. Ibid., p. 38.
11. Ibid., p. 40.
12. Ibid., p. 38.
13. U.S. Congress, Senate Committee on the Judiciary, <u>Soviet Oil in East-West Trade</u>, Hearings Before the Subcommittee to Investigate the Administration of the Internal Security Act and Other Internal Security Laws, 87th Cong., 2d sess. (Washington, D.C.: Government Printing Office, 1962), p. 1.
14. Billerbrook, op. cit., pp. 11-12.
15. Demitri B. Shimkin, <u>Minerals: A Key to Soviet Power</u> (Cambridge: Harvard University Press, 1953), p. 65.
16. U.S. Congress, Senate, Committee on the Judiciary, <u>Soviet Oil in the Cold War</u>, a study prepared by the Library of Congress (by Harold L. Hoskins and Leon Herman) at the Request of the Subcommittee to Investigate the Administration of the Internal Security Act and other Internal Security Laws, 87th Cong., 1st sess. (Washington, D.C.: Government Printing Office, 1961), p. 3.
17. Shimkin, op. cit., p. 66.
18. Ibid.
19. Senate Committee on the Judiciary, <u>Soviet Oil in the Cold War</u>, op. cit., p. 5.
20. Ibid.
21. Ibid., p. 6.
22. Ibid., p. 17.
23. Quotations are from the Soviet publication <u>International Affairs</u>, cited in National Petroleum Council, <u>The Impact of Oil Exports from the Soviet Bloc</u> (Washington, D.C., 1962), p. 42.
24. Ibid., p. A-3.
25. Ibid., p. A-4.
26. Ibid.
27. Ibid., p. A-5.
28. Ibid., p. A-6.
29. Ibid., p. A-8.
30. Ibid., p. A-7.

CHAPTER 9

1. Richard J. Ward, "Soviet Competition in Western Markets: A Commodity Case and Its Implications," <u>Journal of Industrial Economics</u> 8, 2 (March 1960): 147.
2. M. J. Rathbone, president, Standard Oil Company of New Jersey, "The New Interest of American Business,"

a speech before the Tulsa Chamber of Commerce, Tulsa, Oklahoma, December 11, 1956; reprinted in U.S. Congress, Senate Committee on the Judiciary and Committee on the Interior and Insular Affairs, Emergency Oil Lift Program and Related Oil Problems, Joint Hearings (1961), p. 2219.

3.  Remarks by K. S. Adams, chairman and chief executive officer for Philips Petroleum Company, as reported in "Future of Russian Oil," World Petroleum 12, 10 (September 1961): 59.

4.  Figures are from Augustus C. Long, "International Oil--Its Ills and Cures," World Petroleum 12, 1 (January 1961): 37.

5.  Remarks by Secretary of the Interior Stewart L. Udall in a speech before the American Petroleum Institute, July 27, 1961, cited in "Future of Russian Oil," op. cit., p. 59.

6.  Gordon W. Reed, chairman, Texas Gulf Producing Co., "Russia's Oil Drive," World Petroleum 12, 10 (September 1961): 54.

7.  Rathbone, op. cit., p. 2219.

8.  Committee for Economic Development, Soviet Progress vs. American Enterprise (New York, 1958), pp. 30-32..

9.  James Terry Duce, "The Changing Oil Industry," Foreign Affairs 40, 1 (July 1962): 633.

10.  Long, op. cit., p. 55.

11.  Cited in Allen B. Zefros, "Dilemma in Oil," United States Naval Institute Proceedings 90, 2 (February 1964): 61-62.

12.  Clyde la Motte, "The Balance of Payments Continues to Plague U.S.A. Oil Men," World Petroleum Annual Review 37, 8 (July 15, 1966): 8.

13.  Ibid.

14.  Ibid.

15.  "Future of Russian Oil," op. cit., p. 62.

16.  Long, op. cit., p. 56.

17.  Committee for Economic Development, Research and Policy Committee, East-West Trade: A Common Policy for the West, a statement of national policy prepared in association with the European Committee for Economic and Social Progress (CEPES) and Japan Committee for Economic Development (Keizai Daoyukai) (New York, May 1965), p. 38.

18.  Ibid., p. 10.

19.  Ibid.

20.  Ibid., p. 9.

21.  Ibid., p. 39.

22.  William H. Hessler, "Sixth Fleet: Beefed up for a Bigger Job," United States Naval Institute Proceedings 84, 8 (August 1958): 25.

23. Norman Locksley, "NATO's Southern Exposure," United States Naval Institute Proceedings 88, 11 (November 1962): 50.

24. U.S. President (Eisenhower), Adjusting Imports of Petroleum and Petroleum Products to the United States, a proclamation modifying section 2 of the act of July 1, 1954 (72 Stat. 678, 19 U.S.C. 1352a), March 10, 1959, no. 3279; Modifying Proclamation No. 3279 of March 10, 1959, Adjusting Imports of Petroleum and Petroleum Products, a proclamation, April 30, 1959, no. 3290.

25. U.S. Congress, Senate, Committee on Interstate and Foreign Commerce, Ship Transfers to Foreign Flags, Hearings Before the Subcommittee on the Merchant Marine and Fisheries on April 9 and 10, August 20 and 21, 1957, 85th Cong., 1st sess. (Washington, D.C.: Government Printing Office, 1957), 240 pp.

26. Ibid., pp. 1-2.

27. U.S. Congress, Senate, Committee on Interstate and Foreign Commerce, Amendment to the Merchant Marine Act of 1936 (Washington, D.C.: Government Printing Office, 1966), p. 1.

28. James Henry Giffen, The Legal and Practical Aspects of Trade with the Soviet Union (New York: Praeger Publishers, 1971), p. 78.

29. U.S. Congress, House of Representatives, Special Study Mission to Europe, The Soviet Economic Offensive in Western Europe, a report to the Committee of the Whole House on the State of the Union, dated February 7, 1963, 88th Cong., 1st sess. (Washington, D.C.: Government Printing Office, 1963), p. xi.

30. National Petroleum Council, The Impact of Oil Exports from the Soviet Bloc (Washington, D.C., 1962), pp. 43-44, 174-176.

31. Emile Bustani, "Soviet Oil Threatens the Middle East," World Petroleum 12, 2 (February 1961): 41.

32. Ibid., p. 12.

33. Ibid., p. 41.

34. Ibid.

35. Organization of Petroleum Exporting Countries, Public Relations Department, OPEC: Background Information (Vienna, May 1964), p. 5.

36. Ibid.

37. Ibid., p. 10.

38. Francisco R. Parra, "A Role for OPEC," World Petroleum 12, 9 (August 1961): 31.

39. Frank Hendryx, "It's Time for a New Approach to Arab Concession Negotiations," World Petroleum 37, 10 (September 1966): 60.

40. "Russia's Oil Drive," op. cit., p. 56.

41. "Russian Oil Exports--What to Do?" an editorial, World Petroleum 12, 9 (August 1961): 27.

42. Organization for European Economic Cooperation, Europe's Growing Needs of Energy: How Can They Be Met? a report by a group of experts under the chairmanship of Sir Harold Hartley (Paris, May 1956), 120 pp.

43. Organization for European Economic Cooperation, Energy Advisory Commission, Towards a New Energy Pattern in Europe, a report prepared under the chairmanship of Professor Austin Robinson (Paris, January 1960), 125 pp.

44. Organization for European Economic Cooperation, Oil Committee, Oil: The Outlook for Europe, Trends in the Economic Sectors Series (Paris, January 1961), pp. 14, 63, and 64.

45. Ibid., p. 12.

46. Ibid., p. 49.

47. "Petroleum and Natural Gas in Western Europe Today," World Petroleum 35, 6 (June 16, 1964): 37.

48. Ibid., p. 38.

49. Ibid.

50. Paul Hatry, "Future of Oil in the O.E.E.C.," World Petroleum 37, 6 (June 1966): 50 and 52.

51. Ibid., p. 57.

52. John J. Collins, Never Off Pay: The Story of the Independent Tanker Union 1937-1962 (New York: Fordham University Press, 1964), p. 248. "The reason for this is that where transfers of American ships to foreign flags have occurred, generally permission to transfer old ships to these flags has been based on agreement to build newer and faster American flag ships . . . with a carrying capacity double or sometimes triple that of the ships which have been transferred. . . . the Maritime Administration believed that [it was] merely encouraging United States owners of 'war built' vessels to build new and faster vessels to be registered under the U.S. flag" (ibid., pp. 248-249).

53. Ibid., p. 249.

54. Ibid., p. 273.

55. Organization for European Economic Cooperation, Maritime Transport (Paris, 1959), p. 24.

56. Tanker Times 9, 9 (January 1963): 242.

57. L. A. Sawyer and W. H. Mitchell, Merchant Ships of the World: Tankers (New York: Doubleday, 1967), pp. 98 and 100.

58. Ibid., p. 109.

59. Ibid.

60. Preston P. Nibley and Alan C. Nelson, "Economics of Oil Transportation Middle East to Western Europe," World Petroleum 12, 12 (November 1961): 57.

CHAPTER 10

1. U.S. Department of the Interior, The Middle East Petroleum Emergency of 1967 (Washington, D.C., 1969), vol. I, p. 15; vol. II, pp. B-4 and G-27-33.
2. David H. Finnie, Desert Enterprises: The Middle East Oil Industry in Its Local Environment (Cambridge: Harvard University Press, 1958), pp. 142-181.
3. David Hurst, Oil and Public Opinion in the Middle East (London: Faber and Faber, 1966), 127 pp.

CHAPTER 11

1. National Petroleum Council, Future Petroleum Provinces of the United States (Washington, D.C., 1970), p. 107.
2. Ibid., p. 3.
3. Ibid.
4. Richard M. Nixon, "Oil Import Program," A Proclamation Modifying No. 3279, Relating to Imports of Petroleum and Petroleum Products, reprinted in U.S. Department of State Bulletin 67, 1752 (January 22, 1973).
5. Richard M. Nixon, "Oil Import Program," A Proclamation Modifying No. 3279, Relating to Imports of Petroleum and Petroleum Products, reprinted in U.S. Department of State Bulletin 68, 1765 (April 23, 1973).
6. Richard M. Nixon, "Message to Congress," Weekly Compilation of Presidential Documents, April 22, 1973, p. 389.
7. World Petroleum 43, 6 (June/July, 1972): 17.

OFFICIAL SOURCES

Government Publications

Egypt, Iran, Lebanese Republic, Venezuela

Egypt. Ministry of Foreign Affairs. Eastern Mediterranean
and Cyprus. The Tripartite Conference on the Eastern
Mediterranean and Cyprus, London August 29-September
7, 1955, Presented by the Secretary of State for Foreign Affairs to Parliament, Miscellaneous no. 18
(1955), Cmd. 9b94. London: H.M. Stationery Office,
October 1955.

Iran, Government of. "Law Nationalizing the Oil Industry
in Persia," 1 May 1951, reprinted in Cmd. 8425, pp.
29-31 and in Royal Institute of International Affairs,
British Foreign Policy: Some Relevant Documents January 1950-April 1955. London: Oxford University Press,
1955, pp. 11-13.

_____, Iranian Embassy. Some Documents on Nationalization.
Washington, D.C.: Iranian Embassy, n.d.

Lebanese Republic, Government of. Memorandum to the Arab
Council by Emil Bustani. Beirut: American Press,
1959.

Venezuela, Presidencia, Secretaria Géneral. Venezuela and
O.P.E.C. Caracas: Imprenta Nacional, 1961.

United States

(Note: All official U.S. publications are published
by the U.S. Government Printing Office, Washington, D.C.)

United States Bureau of Naval Personnel. Petroleum Logistics. NAVPERS 10892, 1955.

United States Congress. <u>Economic and Military Cooperation</u>
<u>with Nations in the General Area of the Middle East</u>,
Joint Resolution 117, 85th Cong., 1st sess., legisla-
tive day March 2, 1957 (March 5, 1957). Approved by
President Eisenhower on March 9, 1957.

_____, House of Representatives. <u>Message of the Presi-</u>
<u>dent</u> containing a "Declaration of Principles" by the
President of the United States and the Prime Minister
of the United Kingdom, known as the "Atlantic Charter,"
August 14, 1941, House Doct. No. 358, 77th Cong., 1st
sess. (August 21, 1941).

_____, Committee of the Whole House on the State of the
Union. <u>Authorizing the President to Undertake Eco-</u>
<u>nomic and Military Cooperation with Nations in the</u>
<u>General Area of the Middle East</u>, House Report no. 2,
85th Cong., 1st sess. (1957).

_____, Committee on Banking and Currency. <u>Defense Pro-</u>
<u>duction Act Amendments of 1951</u>, Hearings, 82d Cong.,
1st sess. (1951), 3 vols.

_____. <u>Defense Production Act Amendments of 1952</u>, Hear-
ings, 82d Cong., 2d sess. (1952), Part I.

_____. <u>Defense Production Act Amendments of 1952</u>, Report,
House Report no. 2177, 82d Cong., 2d sess. (1952).

_____. <u>Extension of Defense Production Act</u>, Hearings be-
fore Subcommittee no. 1, May 26, 1960, 86th Cong., 2d
sess. (1960).

_____. <u>Extension of Defense Production Act of 1950</u>,
Hearings June 10, 1958, 85th Cong., 2d sess. (1958).

_____. <u>Extension of Defense Production Act of 1950</u>,
Hearings before Subcommittee no. 1, May 9 and 10,
1962, 87th Cong., 2d sess. (1962).

_____. <u>Extension of Defense Production Act of 1950</u>,
Hearing on March 18, 1964, 88th Cong., 2d sess.
(1964).

_____, Committee on Expenditure in Executive Departments.
<u>Petroleum Investigation</u>, Hearings July 1-9, 1947,
80th Cong., 1st sess. (1947).

_____, Committee on Foreign Affairs. Joint Resolution to Promote Peace and Stability in the Middle East: Message from the President, First Report to Congress Covering Activities through June 30, 1957, House Doct. 220, 85th Cong., 1st sess. (August 5, 1957).

_____. The Soviet Economic Offensive in Western Europe, Report of the Special Study Mission to Europe, House Report no. 32, 88th Cong., 1st sess. (1963).

_____, Committee on Interstate and Foreign Commerce. Fuel Investigation: Petroleum and European Recovery Program, Progress Report, 80th Cong., House Documents Series (1948).

_____. Petroleum Supplies for Military and Civilian Needs, final report of the special subcommittee on petroleum investigation, 79th Cong., 2d sess. (1947).

_____. Petroleum Survey, Hearings on the 1957 Oil Outlook, the Oil Lift to Europe and the Price Increases, 85th Cong., 1st sess. (1957).

_____, Committee on Merchant Marine and Fisheries. Controlling Movement and Transportation of Gasoline and Petroleum Products, Report to accompany H.R. 4042, July 21, 1947, 80th Cong. (1947).

_____. Merchant Marine Act of 1936 and the Ship Sales Act of 1946, Hearings, 80th Cong., 2d sess. (1948).

_____. Sale, Charter, and Construction of Vessels, Hearings before the Subcommittee on Merchant Marine, 84th Cong., 2d sess. (1956).

_____. Tanker Trade-in Legislation, Hearings on May 27 and June 3, 1954, 83d Cong., 2d sess. (1954).

_____. To Facilitate and Encourage New Ship Construction, Including National Defense Reserve of Tankers, Hearings on July 23, 24, and 28, 1958 (1958).

_____. Use of American-Owned Tankers Transporting Gasoline and Oil to Russia, Hearings on June 27–July 16, 1947.

_____, Committee on the Judiciary. <u>Iran Consortium Agreement</u>, Hearings before the Antitrust Subcommittee No. 5, 84th Cong. (1955), Text of Agreement, pp. 1563-1651.

_____, House Documents. <u>Program of United States Support for European Recovery</u>, message from President of the United States, December 19, 1947, 80th Cong. (1948).

_____. <u>Supplemental Estimate of Appropriation for Expenses of European Recovery Program, Fiscal Year 1948</u>, April 14, 1948.

_____, House Reports. <u>Foreign Assistance Act of 1948</u>, conference report to accompany S. 2202, April 1, 1948.

_____, Select Committee on Foreign Aid. <u>Petroleum Requirements and Availabilities</u>, Preliminary Report no. 5, 80th Cong. (1947).

_____, Joint Committee on Defense Production. <u>Defense Production Act</u>, Progress Reports and Hearings (various dates).

_____, Senate. <u>Agreement on Petroleum Between the Government of the United States and the Government of Great Britain and Ireland</u>, signed in London on September 24, 1945; agreement and accompanying papers made public on November 2, 1945, Senate Executive Document H, 78th Cong., 2d sess. (1947).

_____. <u>Petroleum Agreement with Great Britain and Northern Ireland</u>, signed in Washington on August 8, 1944; made public on August 25, 1944, Senate Executive Document F, 78th Cong., 2d sess. (1947).

_____. <u>Resolution for a Treaty on the Peaceful Exploration and Exploitation of Ocean Space and its Resources</u> by Senator C. Pell, March 5, 1968, referred to the Committee on Foreign Relations, S. Res. 263, 90th Cong., 2d sess. (1968).

_____, Committee on Banking and Currency. <u>Defense Production Act Amendments of 1951</u>, Hearings, 82d Cong., 1st sess. (1951), 4 vols.

_____. <u>Defense Production Act Amendments of 1955</u>, Hearings before a Subcommittee on June 20, 21, and 27, 1955, 84th Cong., 1st sess. (1955).

264

_____. _Defense Production Act Amendments 1964_, Hearing on June 24, 1964, 88th Cong., 2d sess. (1964).

_____. _Defense Production Act 1962_, Hearings on June 5 and 6, 1962, 87th Cong., 2d sess. (1962).

_____. _To Amend and Extend the Defense Production Act of 1950_, Report to accompany S. 2594, Report no. 1599, 82d Cong., 2d sess. (1952).

_____, Committee on Foreign Relations. _Activities of Nations in Ocean Space_, Hearings before the Subcommittee on Ocean Space, July 24, 25, 28, and 30, 1969, 91st Cong., 1st sess. (1969).

_____. _Anglo-American Oil Agreement_, Report, 79th Cong., 1st sess. (1947).

_____. _European Recovery Program_, basic documents and background information prepared by staffs of Senate Foreign Relations Committee and House Foreign Affairs Committee, November 10, 1948.

_____. _European Recovery Program_, hearings on United States assistance to European economic recovery, January 8-February 5, 1948, 80th Cong., 2d sess. (1948).

_____. _European Recovery Program_, Report (1948).

_____. _Governing the Use of Ocean Space_, Hearing on November 29, 1967, 90th Cong., 1st sess. (1967).

_____. _National Commitments_, S. Rept. 91-129, 91st Cong., 1st sess. (1969).

_____. _Outline of European Recovery Program_, draft legislation and background information submitted by the Department of State for use of Senate Foreign Relations Committee, December 19, 1947 (1948).

_____. _Petroleum Agreement with Great Britain and Northern Ireland_, Hearings, June 2-25, 1947, 80th Cong., 1st sess. (1947).

_____. _The President's Proposal on the Middle East_, Hearings, 85th Cong., 1st sess. (1957).

265

_____, Committee on Government Operations. Sale of
Government-owned Surplus Tanker Vessels, Hearing before the Permanent Subcommittee on Investigations,
82d Cong., 2d sess. (1952).

_____. Sale of Government-owned Surplus Tanker Vessels,
Interim Report by the Permanent Subcommittee on Investigation, Report no. 1613, 82d Cong., 2d sess.
(1952).

_____, Committee on Interior and Insular Affairs. Oil and
Coal Shortages, Hearings before Subcommittee, December
9, 1947, 80th Cong., 1st sess. (1948).

_____, Committee on Interstate and Foreign Commerce.
Amendment to the Merchant Marine Act, 1936, Hearings
before the Subcommittee on Merchant Marine and Fisheries, December 3, 1959, February 15 and 16, 1960,
86th Cong., 1st and 2d sess. (1960).

_____. Cargo Preference Bill, Hearings before a Subcom-
mittee on Water Transportation, 83d Cong., 2d sess.
(1954).

_____. Controlling Movement and Transportation Abroad of
Gasoline and Petroleum Products, Report to accompany
S. 1653, July 23, 1947.

_____. Investigation of Shortage of Petroleum, Petroleum
Products, and Natural Gas, Report, 80th Cong. (1948).

_____. New Ship Construction Program, Hearings before the
Subcommittee on Merchant Marine and Fisheries, 84th
Cong., 1st sess. (1955).

_____. Ship Transfers to Foreign Flags, Hearings before
the Subcommittee on the Merchant Marine and Fisheries
on April 9 and 10, August 20 and 21, 1957, 85th Cong.,
1st sess. (1957).

_____. Tanker and Cargo Tankship Charter and Construc-
tion, Hearing before the Subcommittee on Merchant
Marine and Fisheries, June 13, 1956, 84th Cong., 2d
sess. (1956).

_____, Committee on the Judiciary. Soviet Oil in East-
West Trade, Hearing before the Subcommittee to

266

Investigate the Administration of the Internal Security Act and Other Internal Security Laws, Testimony
of Samuel Nakasian, July 3, 1962, 87th Cong., 2d sess.
(1962).

_____. Soviet Oil in the Cold War, A Study Prepared by
the Library of Congress at the Request of the Subcommittee to Investigate the Administration of the Internal Security Act and Other Internal Security Laws,
87th Cong., 1st sess. (1961).

_____ and Committee on Interior and Insular Affairs.
Emergency Oil Lift Program and Related Oil Problems,
Joint hearings before Subcommittees, 85th Cong., 1st
sess. (1957).

_____, Select Committee on Small Business. The Interna-
tional Petroleum Cartel, Staff Report by the U.S. Federal Trade Commission, submitted to the Subcommittee
on Monopoly, dated August 22, 1952, 82d Cong., 2d
sess. (1952).

_____, Special Committee Investigating National Defense
Program, pt. 41. Petroleum Arrangements with Saudi
Arabia, Hearings, March 28, 1947-January 30, 1948.

_____, Special Committee Investigating Petroleum Resources.
American Petroleum Interests in Foreign Countries,
79th Cong., 1st sess. (1946).

_____. Investigation of Petroleum Resources in Relation
to the National Welfare, Final Report, Report no. 9,
80th Cong., 1st sess. (1947).

_____, Special Committee to Investigate Gasoline and Fuel-
Oil Shortages. Gasoline and Fuel-Oil Shortages, Addi-
tional Report: Interim Report on Oil and Coal, 78th
Cong., 2d sess. (1944).

_____, Senate and House of Representatives, Select Commit-
tees on Small Business. The Third World Petroleum
Congress, A Report by Elmer Patman, 82d Cong., 2d
sess. (1952).

U.S. Department of Commerce, Bureau of the Census. The
Soviet Mineral-Fuel Industries, 1928-1958: A Statistical Survey, International Population Statistics Reports Series P-90, no. 19 (1962).

U.S. Department of the Interior. National Resources and
     Foreign Aid, Report by J. A. Krug, Secretary of the
     Interior, October 9, 1947.

_____, Office of Oil and Gas. The Middle East Petroleum
     Emergency of 1967 by Wilson M. Laird, director (1969).

_____. Report to the Secretary of the Interior from the
     Director of the Voluntary Agreement Relating to For-
     eign Petroleum Supply, as amended May 8, 1956, con-
     cerning the activities of the Foreign Petroleum Sup-
     ply Committee under the Voluntary Agreement and the
     activities of the Middle East Emergency Committee and
     its Subcommittees under the Plan of Action, for the
     period of April 1, 1956 through June 30, 1957 (mimeo-
     graphed, 1957).

U.S. Department of State, Public Affairs Office, Publica-
     tions Division. Assistance to European Economic Re-
     covery, statement by Secretary George C. Marshall,
     Secretary of State, before Senate Committee on Foreign
     Relations, January 8, 1948, and the President's mes-
     sage to Congress, December 19, 1947, Economic Coopera-
     tion series, Publication 3022 (1948).

_____. European Recovery Program: Commodity Reports in-
     cluding Manpower, Economic Cooperation series, Publi-
     cation 3093 (1948).

U.S. Library of Congress, Legislature Reference Service.
     Problems Raised by the Soviet Union Oil Offensive,
     study prepared for the Subcommittee to Investigate
     the Administration of the Internal Security Act and
     Other Internal Security Laws of the Committee of the
     Judiciary of the U.S. Senate (1962).

U.S. President (Eisenhower). Modifying Proclamation No.
     3279 of March 10, 1959, Adjusting Imports of Petroleum
     and Petroleum Products, a proclamation, April 30, 1959,
     No. 3290.

_____. Program for the Middle East, an address before the
     United Nations General Assembly August 13, 1958, De-
     partment of State Publication 6697, International Or-
     ganization and Conference series III, 130.

_____. The Situation in the Middle East (Radio and Tele-
     vision Address to the American People, February 20,

1957), Department of State Publication 6461, Near and Middle Eastern Series 23.

_____, Special Committee to Investigate Crude Oil Imports. Report, dated July 29, 1957.

U.S. President (Truman). Message to Congress: Defense Production Act, February 11, 1952, Document no. 347 of the House of Representatives, 82d Cong., 2d sess. (1952).

_____, Office of Government Reports. European Recovery and American Aid, report by the President's Committee on Foreign Aid, November 7, 1947.

U.S. Public Laws. Act to Promote World Peace and General Welfare, National Interest, and Foreign Policy of United States through Economic, Financial, and other measures necessary to maintenance of conditions abroad in which Free Institutions may survive and consistent with maintenance of strength and stability of United States, approved April 3, 1948, 80th Cong., 2d sess.

U.S. Statutes at Large. An Act to Promote Export Trade, and for Other Purposes, H.R. 2316, April 10, 1918, Volume 40 (1919).

_____. Defense Production Act of 1950, Volume 64, Part I: Public Laws and Reorganization Plans, H.R. 9176, Public Law 774 (1952), pp. 798-822; Section 708: pp. 818-19.


International Organizations Publications

Organization for Economic Cooperation and Development, Special Committee for Oil. Oil Today. Paris, 1964.

Organization for European Economic Cooperation. Basic Statistics of Energy for OEEC Countries 1950-1958. Paris, 1959.

_____. Europe's Growing Needs of Energy: How Can They Be Met? A Report prepared by a Group of Experts. Paris, May 1956.

_____. Europe's Need for Oil: Implications and Lessons of the Suez Crisis. Paris, January 1958.

_____. _The Search for and Exploitation of Crude Oil and Natural Gas in the O.E.E.C. Area_. Paris, December 1957.

_____. _Towards a New Energy Pattern in Europe_. Paris, January 1960.

_____, Maritime Transport Committee. _Maritime Transport_. Paris, annual.

_____, Oil Committee. _Oil: The Outlook for Europe_, Doct. No. PE(56)2. Paris, September 1956.

_____. _Oil: Recent Developments in the O.E.E.C. Area_, Doct. No. C(60)214. Paris, January 1961.

_____. _Oil Refinery Expansion_. Paris, October 1949, August 1951, June 1953.

Organization of Petroleum Exporting Countries, Public Relations Department. _Background Information_. Vienna, 1964, c. 1966.

_____. _The Statute of the Organization of the Petroleum Exporting Countries_. Vienna, n.d.

Permanent Council of the World Petroleum Congress. _Suez Company Warns World Economy Hangs on Canal Expansion_. New York, press release, April 3, 1957.

United Nations, Economic and Social Council. _Control of Oil Resources_, item proposed by the International Co-operative Alliance, E/449, mimeographed, 2 July 1947.

_____. _Control of World Oil Resources_, item 26, a proposal for the creation of a United Nations Petroleum Commission under the authority of the Economic and Social Council, presented by the International Co-operative Alliance, supporting document No. 2, E/449/ Add. 1, mimeographed 31 July 1947.

_____, Economic Commission for Europe. _Energy Situation in Europe_. New York: World Mart Publishers, 1958.

National Organizations Publications

American Petroleum Institute. Petroleum Facts and Figures: 1967. New York, December 1967.

Committee for Economic Development. Soviet Progress vs. American Enterprise, Report of a confidential briefing session held at the fifteenth anniversary meeting of the CED on November 21, 1957, in Washington, D.C. Garden City, N.Y.: Doubleday, 1958.

_____, Research and Policy Committee. East-West Trade: A Common Policy for the West. New York, May 1965.

_____. The European Common Market and Its Meaning to the United States. New York, 1959.

_____. Japan in the Free World Economy. New York, March 1963.

_____. A New Trade Policy for the United States. New York, April 1962.

_____. Trade Negotiations for a Better Free World Economy. New York, 1964.

Council on Foreign Relations. Documents on American Foreign Relations. New York: Harper, annual.

Foreign Petroleum Supply Committee. Report to the Petroleum Administration for Defense concerning the organization and activities of the committee and its subcommittees during the period ending December 31, 1951 (1952).

_____. Report to the Secretary of the Interior and the Petroleum Administrator for Defense on the voluntary aid program pursuant to the Voluntary Agreement dated June 25, 1951, and Plan of Act No. 1 dated July 26, 1951, for the period ending December 31, 1951 (1952).

Institute for Strategic Studies. The Military Balance. London, annual.

National Petroleum Council. A National Oil Policy for the United States. Washington, D.C., 1949.

_____, Committee on National Petroleum Emergency.  <u>Report</u>.
Washington, D.C., January 13, 1949.

_____, Committee on Petroleum Imports.  "Report."  Wash-
ington, D.C., May 5, 1955.

_____, Committee on Tanker Requirements.  "Report."
Washington, D.C., March 7, 1957.

_____, "World Petroleum Tanker Construction."  Washing-
ton, D.C., March 7, 1957.

_____, Committee on the Impact of Oil Exports from the
Soviet Bloc.  <u>Impact of Oil Exports from the Soviet
Bloc</u>, A Report dated October 4, 1962.  Washington,
D.C., 1962.

_____.  <u>Impact of Oil Exports from the Soviet Bloc, Sup-
plementary</u>, a revision of the report issued October
4, 1962, adopted by the NPC March 19, 1964.  Washington, D.C., 1964.

Royal Institute of National Affairs.  <u>Survey of International Affairs</u>.  London:  Oxford University Press,
various dates.

_____, Information Department.  "The Baghdad Pact:  Ori-
gins and Political Setting."  London:  Chatham House,
February 1956.

_____.  "Current Soviet Policies," an appraisal of the
20th Congress of the Communist Party of the Soviet
Union.  London, March 1956.

_____.  "The Organization for European Economic Co-
operation."  London, November 1958.

_____.  "The Western Powers and the Middle East 1959."
London, 1958, revised June 1959.


SECONDARY SOURCES

Books and Theses

Abu el-Haj, Ribhi.  "Oil Industry:  A Strategic Factor in
the Development of Iraq," thesis, Columbia University.
Ann Arbor, Mich.:  University Microfilms AC-1, no.
58-1328, 1958.

Adams, Michael. Chaos or Rebirth: The Arab Outlook.
    London: BBC, 1968.

Ady, Peter, ed. Private Foreign Investment and the De-
    veloping World. New York: Praeger Publishers, 1971.

Anderson, Ronald A. Government Regulation of Business.
    Cincinnati: South-Western Publishing Company, c. 1966.

Athay, Robert E. The Economics of Soviet Merchant Shipping
    Policy. Chapel Hill: University of North Carolina
    Press, 1971.

Badeau, John S. American Approach to the Arab World. New
    York: Council on Foreign Relations, Harper and Row,
    1968.

Al-Baharana, Hussain M. The Legal Status of the Arabian
    Gulf States. Manchester: Manchester University
    Press, 1966.

Baldwin, George B. Planning and Development in Iran.
    Baltimore: Johns Hopkins Press, 1967.

Barrows, Gordon Hensley. International Petroleum Industry.
    New York: International Petroleum Institute, 1965.

Becker, Abraham S. Oil and the Persian Gulf in Soviet
    Policy in the 1970's. Santa Monica, Cal.: RAND
    Corporation, December 1971.

Bergier, Jacques and Bernard Thomas. La guerre secrète du
    pétrole. Paris: Editions Deoel, 1968.

Billerbeck, Klaus. "Soviet Bloc Foreign Aid to the Under-
    developed Countries." Hamburg: Hamburg Archives of
    World Economy, 1960.

Brodie, Bernard. "Foreign Oil and American Security,"
    memorandum no. 23. New Haven, Conn.: Yale Institute
    of International Studies, September 15, 1947.

Bullard, Sir Reader, ed. The Middle East: A Political
    and Economic Survey. London: Oxford University Press
    for the Royal Institute of International Affairs, 3d
    ed., 1958.

Campbell, Robert W. _Soviet Economic Power, Its Organization, Growth, and Challenge_. Boston:  Houghton Miff-
lin, 1960; 2d ed., 1966.

Carmical, J. H.  _The Anglo-American Petroleum Pact_. New
York:  economic survey series No. 408, American Enterprise Association, 1945.

Chapelle, Jean.  _Géographie économique du pétrole_, publications de l'Institut français du pétrole.  Paris:
Technip, 1968.

_____, and Sonia Ketchian.  _URSS, second producteur de
pétrole du monde_, publications de l'Institut français
du pétrole.  Paris:  Technip, 1963.

De Chazeau, Melvin Gardner, and Alfred E. Kahn.  _Integra-
tion and Competition in the Petroleum Industry_.  New
Haven, Conn.:  Yale University Press, 1959.

Duncan, W. Raymond, ed.  _Soviet Policy in Developing Coun-
tries_.  Waltham, Mass.:  Ginn-Blaisdell, 1970.

Durand, Daniel.  _La politique pétrolière internationale_.
Paris:  Presses universitaires de France, 1960.

Eden, Sir Anthony.  _Memoirs:  Full Circle_.  London:
Cassell, 1960.

Eisenhower, Dwight D.  _The White House Years_, vol. I:
"Waging Peace 1956-1961."  New York:  Doubleday, 1965.

Engler, Robert.  _The Politics of Oil:  A Study of Private
Power and Democratic Directions_.  New York:  Macmillan,
1961; Chicago:  Phoenix Books, The University of
Chicago Press, 1967.

Fanning, Leonard M.  _Foreign Oil and the Free World_.  New
York:  McGraw-Hill, 1954.

_____.  "The Shift of World Petroleum Power Away from the
United States."  Pittsburgh:  American Petroleum Institute, April 10, 1958.

Feis, Herbert.  _Petroleum and American Foreign Policy_, Food
Research Institute of Stanford University, Commodities
Policy Study No. 3.  Stanford University, Cal., March
1944.

_____. _Seen from E.A._  New York:  Alfred A. Knopf, 1947.

Finer, Herman.  _Dulles over Suez:  The Theory and Practice of His Diplomacy_.  Chicago:  Quadrangle Books, 1964.

Ford, Alan W.  _The Anglo-Iranian Oil Dispute of 1951-1952: A Study of the Role of Law in the Relations of States_.  Berkeley and Los Angeles:  University of California Press, 1954.

Frank, Helmut J.  _Crude Oil Prices in the Middle East_.  New York:  Praeger Publishers, 1966.

Frankel, P. H.  _Essentials of Petroleum:  A Key to Oil Economics_.  New York:  Augustus M. Kelley Publishers, 1969.

_____.  _Mattei:  Oil and Power Politics_.  London:  Faber and Faber, 1966.

Gannett, Ernest.  _Tanker Performance and Cost_.  Cambridge, Md.:  Corvell Maritime Press, 1969.

Giffen, James Henry.  _The Legal and Practical Aspects of Trade with the Soviet Union_.  New York:  Praeger Publishers, October 1971.

Gullion, Edmund A., ed.  _Uses of the Seas_.  Englewood Cliffs, N.J.:  Prentice-Hall, for The American Assembly, Columbia University, 1968.

Guyol, N. B.  "Free World Energy Consumption by Area and Product," internal memorandum.  San Francisco:  Economics Department of the Standard Oil Company of California, May 1958.

Halpern, Manfred.  _The Politics of Social Change in the Middle East and North Africa_.  Princeton, N.J.:  Princeton University Press, 1963.

Hartshorn, J. E.  _Oil Companies and Governments:  An Account of the International Oil Industry in Its Political Environment_.  London:  Faber and Faber, 1962; published in the United States under the title _Politics and World Oil Economics_.  New York:  Praeger Publishers, 2d rev. ed., 1967.

Hirst, David. _Oil and Public Opinion in the Middle East_.
  London:  Faber and Faber, 1966.

Hodkins, Jordan A.  _Soviet Power:  Energy Resources, Production and Potentials_.  Englewood Cliffs, N.J.:
  Prentice-Hall, 1961.

Hoskins, Halford L.  _Middle East Oil in United States Foreign Policy_, by the Legislative Reference Service of
  the Library of Congress, Public Affairs Bulletin no.
  89.  Washington, D.C.:  Government Printing Office,
  1950.

Huie, William O., Marvin Kenneth Woodward, and Ernest E.
  Smith, III.  _Cases and Materials on Oil and Gas_.  St.
  Paul, Minn.:  West Publishing Co., 1972.

Hunter, Robert.  _The Dilemma of Soviet Involvement in the
  Middle East_.  London:  Institute of Strategic Studies,
  1969/1970.

_____.  _Security in Europe_.  London:  Elek Books, 1969.

Hurewitz, Jacob C.  _Middle East Dilemmas:  The Background
  of United States Policy_.  New York:  Harper for the
  Council on Foreign Relations, 1953.

_____.  _Middle East Politics:  The Military Dimensions_.
  London:  Pall Mall for the Council on Foreign Relations, 1969.

_____.  _Soviet-American Rivalry in the Middle East_, papers
  presented at a Conference sponsored by the Ford Foundation on December 13 and 14, 1968, at Columbia University.  New York:  Academy of Political Science, Columbia University, 1969.

Ickes, Harold L.  _Fighting Oil_.  New York:  Alfred A.
  Knopf, 1943.

_____.  _An Oil Policy_, an open letter to the Members of
  the Congress of the United States.  Washington, D.C.:
  Government Printing Office, May 30, 1947.

Ismail, Salem K.  _The Correlation Between Energy Consumption and Gross National Product_.  Vienna:  OPEC, 1968.

Issawi, Charles, and Mohammed Yaganeh. The Economics of
    Middle East Oil. New York: Praeger Publishers, 1962.

Laqueur, Walter Z. The Struggle for the Middle East: The
    Soviet Union and the Middle East 1958-68. London:
    Routledge and Kegan Paul, 1969.

Leeman, Wayne A. The Price of Middle East Oil: An Essay
    in Political Economy. Ithaca, N.Y.: Cornell Uni-
    versity Press, 1962.

Lenczowski, George. The Middle East in World Affairs.
    Ithaca, N.Y.: Cornell University Press, 1962.

_____. Oil and State in the Middle East. Ithaca, N.Y.:
    Cornell University Press, 1960.

_____. The Political Awakening of the Middle East.
    Englewood Cliffs, N.J.: Prentice-Hall, 1970.

_____. Russia and the West in Iran 1918-1948: A Study
    in Big-Power Rivalry. Ithaca, N.Y.: Cornell Uni-
    versity Press, 1949; New York: Greenwood Press, 1968.

_____. Soviet Advances in the Middle East. Washington,
    D.C.: American Enterprise Institute, 1946.

Lewis, Bernard. The Middle East and the West. London:
    Weisenfeld Goldbacks, c. 1968.

Longrigg, Stephen Hemsley. Oil in the Middle East: Its
    Discovery and Development. London: Oxford Univer-
    sity Press, 1961.

Lubell, Harold. Middle East Oil Crises and Western Europe's
    Energy Supply. Baltimore: Johns Hopkins Press for
    RAND Corporation, 1963.

_____. "Middle East Crises and World Petroleum Movements."
    Santa Monica, Cal.: Research Memorandum RM-2185 ASTIA
    Document No. AD156018, RAND Corporation, 26 May 1958.

_____. "The Soviet Oil Offensive and Inter-bloc Economic
    Competition." Santa Monica, Cal.: Memorandum RM-
    2812-PR, RAND Corporation, December 1961.

Lufti, Ashraf T. Arab Oil: A Plan for the Future. Beirut:
    Middle East Research and Publishing Center, 1960.

_____. O.P.E.C. Oil. Beirut:   Middle East Research and
      Publishing Center, 1968.

Marshall, George Catlett.  Assistance to European Economic
      Recovery, statement before Senate Committee on Foreign
      Relations, January 8, 1948, Department of State Publi-
      cation 3022, Economic Cooperation series, 2.  Washing-
      ton, D.C.:  Government Printing Office, 1948.

Paciti, Sebastian.  The Oil Import Problem.  New York:
      Fordham University Press, 1958.

Rouhani, Fuad.  A History of O.P.E.C.  New York:   Praeger
      Publishers, November 1971.

Schurr, Sam H.  Foreign Trade Policies Affecting Mineral
      Fuels in the United States and Western Europe.  New
      York:  paper presented to the American Association
      for the Advancement of Science, December 30, 1960.

_____, Paul T. Homan, et al.  Middle Eastern Oil and the
      Western World:  Prospects and Problems.  New York:
      American Elsevier, 1971.

_____, Bruce C. Natschet, et al.  Energy in the American
      Economy 1850-1975:  An Economic Study of Its History
      and Prospects.  Baltimore:  Resources for the Future,
      Johns Hopkins Press, 1960.

Scott, John C.  Antitrust and Trade Regulations Today:
      1969.  Washington, D.C.:  Bureau of National Affairs,
      1969.

Shwadran, Benjamin.  The Middle East, Oil and the Great
      Powers.  New York:  Praeger Publishers, 1955.

Stocking, George W.  Middle East Oil:  A Study in Political
      and Economic Controversy.  Kingsport, Tenn.:   Vander-
      bilt University Press, 1970.

_____, and Myron W. Watkins.  Monopoly and Free Enter-
      prise, with the report and recommendations of the
      Committee on Cartels and Monopoly.  New York:  Green-
      wood Press, 1968.

Tugendhat, Christopher.  Oil:  The Biggest Business.  New
      York:  G. P. Putnam's Sons, 1968.

Uri, Pierre. _Premier rapport sur une politique coordinée
dans le domaine de l'énergie_. Luxemburg:  European
Coal and Steel Community, April 1959.

Van Den Heuvel, J. A.  "Integrated Energy for Europe,"
paper prepared for the Round Table Conference on
European Problems.  London, November 3 and 4, 1959.

Vansant, Carl.  _Strategic Energy Supply and National Se-
curity_.  New York:  Praeger Publishers, November 1971.

Votaw, Dow.  _The Six-Legged Dog:  Mattei and ENI:  A Study
in Power_.  Berkeley:  University of California Press,
1964.

Williams, Howard R., and Charles J. Meyers.  _Oil and Gas
Law_.  New York:  Matthew Bender and Company, 1959,
4th ed., 1971, 7 vols.

Zimmerman, Erich W.  _Conservation in the Production of
Petroleum_, Petroleum Monograph Series Volume 2,
American Petroleum Institute.  New Haven, Conn.:
Yale University Press, 1957.

Berlin, 183
Biddle, Francis, 88
Black Sea, 173
Bonaparte, Napoleon, 123
Brewster, Ralph Owen, 38
British Broadcasting Corporation (BBC), 83
British Channel, 123
British Dominions, 104
British Empire, 7, 8, 27
British Isles. See Great Britain
British Petroleum Company (BP) Ltd., 95, 121
British Petroleum Tanker Company Ltd., 208
British Shell Oil Company, 34
Bulganin, Nikolai Alexandrovich, 126
Bureau of Mines, U.S., 132, 220
Burma, 121
Burmah Oil Company Ltd., 81
Byrnes, James Francis, 35, 77

Cairo Conference, 42
California, 4
California-Texas Oil Company (Caltex), 10
Cambodia, 228
Canada, 9, 105, 215, 220, 230
Cape of Good Hope, 122, 124, 127, 216, 217
Caribbean, 8, 23, 90, 112, 118, 128, 130
Central America, 4, 88
Central Powers, 131
Central Treaty Organization (CENTO), 147
Ceylon, 177
Chase Manhattan Bank, 89
China, 68, 119, 170
Churchill, Sir Winston S., 81

Coast Guard, U.S., 131
Codman, Sir John, 7
Coleman, Stewart P., 89, 138, 140
Cold War, 56, 58, 59, 77, 82, 96, 123, 133, 145, 146, 158, 162, 169, 197
Colonial Office (Great Britain), 9
Committee for Economic Development (CED), 189, 192-93
Committee of Oil Industrialists, 39, 47, 49
Common Market Commission, 197
Communist International (Comintern), 159, 160
Communist Party of Iran (CPI), 34. See also Tudeh Party
Communist Party of the Soviet Union (CPSU), 182; Twentieth Congress of, 182
Compagnie Française des Pétroles, 8, 95
Congress, U.S., 12, 32, 72, 99, 106, 133, 134, 135, 137, 148, 193, 196, 197, 226, 230, 237
Connally, Tom, 46
Continental Oil Company, 215
Coolidge, J. Calvin, 8
Costa Rica, 111
Crowley, Leon T., 22
Cuba, 172
Curtice, A. A., 34
Curzon, Lord, 6
Cyprus, 194
Czechoslovakia, 119

D'Arcy, William Knox, 81
Defense Production Act, 64-65, 89, 118, 131, 133-35, 136, 137, 143
Defense Production Agency, 89
Denmark, 104, 105, 110, 111, 168, 175

Germany, 76, 114, 115, 175, 181. See also Imperial German Government
Getty, George F., II, 189, 190
Ghana, 145, 179, 180
Gladstone, William Ewart, 124
Glubenkian, C. S., 8
Godber, Sir Frederick, 47
Grady, Henry F., 85, 86
Great Britain, 3, 5, 6, 7, 9, 27, 36, 37, 38, 42, 43, 44, 45, 46, 47, 48, 53, 58, 67, 75, 79, 80, 81, 82, 83, 85, 86, 88, 90, 91, 92, 93, 104, 105, 107, 110, 114, 115, 119, 121, 122, 123, 124, 125, 146, 147, 152, 153, 155, 168. See also Colonial Office; Foreign Office; Labour Government; Royal Navy
Greece, 59, 79, 82, 98, 150, 168, 169, 170, 193, 207
Guatemala, 183
Gulf Oil Corporation, 9, 23, 95
Haifa, 26, 43
Halim Pasha, Grand Vizir Said, 5
Hancock Oil Company, 96
Harden, Orville, 40
Hare, Ambassador Raymond A., 15
Harriman, Averell, 72, 93
Harriman Report on European Recovery and American Aid, 72
Hartley, Sir Harold, 202
Harvey, George, 6
Henderson, Loy W., 94
Herter, Christian A., 72
Herter Report on Petroleum Requirements and Availabilities, 72
Hill, George A., 28
Hitler, Adolf, 161

Homs (Syria), 178
Honduras, 111
Hoover, Herbert, 6
Hoover, Herbert Jr., 34, 93-94
House of Representatives (U.S. Congress), 116
House Select Committee on Foreign Aid, 72
Hughes, Charles Evans, Secretary of State, 8
Hull, Cordell, 22
Hurley, Brigadier General Patrick J., 21
Hungary, 161

Ibn Saud, King, 20-22, 31
Ickes, Harold L., 12, 17, 19-29, 39-40, 47, 48
Imperial German Government, 67
Independent Petroleum Association of America, 28
India, 92, 145, 168, 172, 173, 177, 180
Indian Ocean, 18, 80, 177, 180, 194
Indonesia, 145, 177, 180, 200
International Bank for Reconstruction and Development, 80
International Cooperative Alliance, 34, 57, 58, 59
International Court of Justice, 86
International Petroleum Commission, 44
International Oil Congress at Paris (1933), 11
Interstate Oil Compact, 12, 13, 40, 41
Iowa, 227
Iran (Persia), 34-37, 38, 43, 45, 58, 59, 76, 78, 79-99, 107, 112, 121, 135, 147, 150, 177-78, 183, 200, 219, 238

Iranian Oil Consortium,
93-96, 97
Iranian Oil Exploration and
Producing Company, 94-95
Iraq, 3, 5, 7, 8, 9, 10, 43,
59, 95, 121, 125, 147, 178,
200, 219, 234
Iraq Petroleum Company (IPC),
9, 122, 127, 129, 138, 234
Israel, 118, 122, 170
Italian colonies, 161, 193
Italy, 68, 75, 88, 92, 104,
105, 110, 114, 115, 161,
169, 172, 175

Jackson, Henry M., 235
Japan, 68, 108, 110, 112,
168, 170, 173, 175, 198,
208
Johnson, Lyndon B., 190
Jordan, 125, 219
Jowitt, Viscount, 83

Kabul (Afghanistan), 180
Kefauver, Estes, 142
Kendall, Charles H., 136
Khrushchev, Nikita Sergeye-
vich, 149, 157, 163, 166,
167, 169, 181, 182, 183,
184
Kirkuk-Tripoli Pipeline, 127
Knox, W. Franklin, 22
Korea (including Korean
War), 64, 86, 134, 138,
144, 183, 188, 229
Korea, North, 170
Krug, A. J., 61, 72, 116,
138
Krug Report on National Re-
sources and Foreign Aid,
72
Kuwait, 9, 23, 24, 26, 67,
90, 219
Kuwait Neutral Zone, 9

Labour Government, British,
48
Latin America, 174

League of Nations, 68, 76
League of Nations Council, 7
Lebanon, 37, 219
Lenczowski, George, 43
Lend Lease, 20, 21, 22, 31,
56
Lend Lease Administration,
21
Lenin, Vladimir Ilych
Ulyanov, 36, 159, 163,
182, 183
Lesseps, Ferdinand de, 124
Liberia, 110, 111, 207
Libya, 200, 219
Lodge, Henry Cabot, 5, 38
London, 38, 42-43, 81
London Conference on the
Suez Canal, 151
Long, A. C., 89

Macmillan, Harold, 148
Magnuson, Warren G., 196
Maine, 38
Majlis, 34, 81, 82, 84, 85
Mali, 179, 180
Marathon Oil Company, 215
Maritime Commission, U.S.,
106
Marshall Plan, 73, 74, 105,
113, 128, 133, 149, 150,
152
Massachusetts, 38-39
Mediterranean Sea, 18, 24,
26, 28, 38, 53, 81, 121,
124, 127, 168, 169-70,
179, 186, 193, 194, 217,
218
Merchant marine, U.S., 102,
103, 105, 106, 107, 110,
116
Merchant Marine Act, 103,
107, 116, 196
Mexico, 4, 23, 43, 45
Middle East Emergency Com-
mittee (MEEC), 64, 118,
131, 135, 137-45, 211
Military Sea Transportation
Service, 106, 107, 131

Sixth Fleet, U.S., 194
Socony-Vacuum Oil Company,
   55, 60, 95
South America, 4, 88, 121
South-East Asia Treaty Or-
   ganization (SEATO), 125
Soviet bloc, 121, 123, 128,
   175, 176, 181, 192, 193,
   197, 198, 204
Soviet-Iranian Stock Com-
   pany, 177
Soviet Union, see USSR
Spain, 105, 145, 168, 169
Sputnik, 183
Stalin, Joseph, 157, 159,
   160, 162, 163, 182, 183
Standard Oil Company of
   California, 9, 10, 25,
   93-94, 95
Standard Oil Company of New
   Jersey, 4, 5, 6, 40, 89,
   95, 130, 138, 173, 188
Standard Oil Company of
   Ohio, 96
Standard Vacuum Oil Company,
   34
Stettinius, Edward R. Jr.,
   21, 42, 44
Stewart, Hugh, 138
Strathcona, Lord, 81
Suez Canal, 25, 29, 92, 109,
   112, 114, 117-53, 157, 161,
   183, 195, 202, 206, 208,
   209, 211, 216, 217, 218,
   220, 221, 229
Suez Canal Maritime Company,
   151.  See also Universal
   Maritime Canal Company.
Suez Canal Users' Associa-
   tion, 151
Suez Canal Zone, 120, 147
Suez Gulf, 179
Sukarno, 145
Supreme Court, U.S., 12, 15,
   55, 60-61
Sweden, 104, 105, 110, 111,
   168, 170, 175
Syria, 5, 37, 122, 125, 127,
   138, 169, 178

Teheran, 34, 83, 84, 85
Teheran Conference, 42
Texas, 4, 8, 46
Texas Oil Company, 10, 25,
   89, 95
Thames, 124
Third World, 125, 145, 146,
   151, 159
Tibet, 36
Tidewater Oil Company, 96,
   189
Tito, Josip, 145
Trade Expansion Act (1962),
   196
Trieste, 193
Tripartite Declaration, 146,
   148
Truman, Harry S., 49, 93,
   105, 134
Truman Doctrine, 59, 79, 98,
   150
Tudeh Party, 35, 79, 80, 83,
   150.  See also Communist
   Party of Iran (CPI)
Turkey, 6, 7, 79, 147, 150
Turkish Empire, 5
Turkish Petroleum Company,
   5-6, 7, 8

Union of Soviet Socialist
   Republics (USSR), 27, 34,
   35, 36, 37, 38, 42, 43,
   45, 46, 56, 57, 58, 59,
   79, 80, 82, 83, 84, 85,
   86, 98, 119, 120, 123,
   124, 125, 126, 146, 149,
   150, 151, 153, 157-84,
   187, 192, 193, 198, 199,
   201, 202, 204, 206, 210,
   212, 213, 234, 239
United Geological Company,
   93
United Kingdom.  See Great
   Britain
United Nations, 34, 84, 220;
   UN Afro-Asian Members, 125;
   UN Economic and Social Council, 56-57; UN General Assembly, 125, 145, 146; UN
   Security Council, 123

United States Geological
   Survey, 4
Universal Maritime Canal
   Company, 119.  See also
   Suez Canal Maritime Com-
   pany

Vargas, Eugene, 157
Venezuela, 82, 84, 172,
   185, 198, 200
Versailles Peace Conference,
   39
Vickers-Armstrong (Ship-
   builders) Ltd., 207
Vietnam, 228
Vietnam, North, 170
Voice of America, 83
Voluntary Agreements Relat-
   ing to Foreign Petroleum
   Supply, 34, 63, 64-66,
   118, 129, 131, 135,
   136-37, 138, 141
Voluntary Balance of Pay-
   ments, 190

War Department, 52, 53, 54.
   See also Department of
   Defense; Pentagon
War Production Board, 63
Warren, H. Ed., 132
Warsaw Pact, 123, 125
Washington, 6, 35, 38, 42,
   43, 46, 47, 56, 72, 79,
   84, 93, 96, 97, 107, 132,
   151, 153, 193, 194, 212
Washington Conference, 45
Webb-Pomerene Act, 61
Western Alliance.  See
   North Atlantic Treaty
   Organization
Western Hemisphere, 52, 71,
   87, 99, 121, 128, 130,
   168, 204, 225, 232, 233
White, Dr. David, 4
Wilbur, Ray Lyman, 132
World Bank, 119
World Revolution, 119

World War I, 4, 13, 41, 67,
   76, 131, 133, 159
World War II, 15, 16, 17,
   18, 30, 31, 38, 67, 76,
   78, 79, 97, 102, 111, 117,
   124, 133, 138, 144, 145,
   146, 150, 183, 216, 229
Wormser, Felix E., 128-29

Yemen, 179
Yugoslavia, 145, 170, 181

**RELATED TITLES**
Published by
Praeger Special Studies

THE PRICING OF CRUDE OIL:  Economic
and Strategic Guidelines for an
International Energy Policy
                    Taki Rifaï


A HISTORY OF O.P.E.C.
                    Fuad Rouhani


REGULATING THE NATURAL GAS INDUSTRY:
Pipelines, Conglomerates, and
Producers
                    Samuel Blitman


ISRAEL AND IRAN:  Bilateral
Relationships and Effect on the
Indian Ocean Basin
                    Robert B. Reppa, Sr.